纺织艺术设计
TEXTILE DESIGN

2012年第十二届全国纺织品设计大赛暨国际理论研讨会
12TH CHINA TEXTILE DESIGN COMPETITION & INTERNATIONAL CONFERENCE 2012

2012年国际植物染艺术设计大展暨理论研讨会——传承与创新
INTERNATIONAL PLANT DYEING ART EXHIBITION & CONFERENCE—INHERITANCE & INNOVATION 2012

国际植物染作品集
WORKS COLLECTION OF INTERNATIONAL PLANT DYEING

田青 主编

清华大学美术学院
2012年第十二届全国纺织品设计大赛暨国际理论研讨会组委会 编

中国建筑工业出版社

图书在版编目（CIP）数据

纺织艺术设计　2012年第十二届全国纺织品设计大赛暨国际理论研讨会　2012年国际植物染艺术设计大展暨理论研讨会——传承与创新　国际植物染作品集/田青主编.—北京：中国建筑工业出版社，2012.3
ISBN 978-7-112-14071-8

I.①纺…　II.①田…　III.①纺织品-染整-图集　IV.①TS190.6-64

中国版本图书馆CIP数据核字（2012）第030255号

责任编辑：吴　绫　李东禧
责任校对：党　蕾　陈晶晶

纺织艺术设计
2012年第十二届全国纺织品设计大赛暨国际理论研讨会
2012年国际植物染艺术设计大展暨理论研讨会——传承与创新
国际植物染作品集
田青　主编
清华大学美术学院
2012年第十二届全国纺织品设计大赛暨国际理论研讨会组委会　编

*

中国建筑工业出版社出版、发行（北京西郊百万庄）
各地新华书店、建筑书店经销
北京嘉泰利德公司制版
北京中科印刷有限公司印刷

*

开本：880×1230毫米　1/16　印张：13$\frac{1}{4}$　字数：410千字
2012年3月第一版　2012年3月第一次印刷
定价：**126.00元**
ISBN 978-7-112-14071-8
　　（22119）

版权所有　翻印必究
如有印装质量问题，可寄本社退换
（邮政编码 100037）

卷首语

　　今年是"全国纺织品设计大赛暨国际理论研讨会"举办的第12年。清华大学美术学院作为此项活动的创始者与承办者,始终秉承建院伊始就确立的艺术设计要面向"衣、食、住、行、用"的学科建设与人才培养理念,坚持大赛及论坛染织与服装专业指向。因此得到国内外高校及同业的高度认可,创办至今已吸引了国内外数十所艺术和综合类高校的数千名师生的积极参与,通过不同的专业主题设计充分体现各高校的专业特色与艺术设计水平。同时,我们在每届成果的基础上,与时俱进,积极调整,使大赛主题在传承传统的同时又具有鲜明的时代特征。

　　今年,大赛承袭"上善若水"之宗旨,以"人与自然和谐相处、人与人爱心互助的可持续发展"为主线,以"纺织品的天然植物染"作为国际纺织艺术展览与国际理论研讨会的主题词。来自国内22所高校和国外21所高校的师生,以及国内外19所纺织研究机构共同参与了此次艺术盛会与学术交流论坛。众多参赛作品通过深邃的立意与创造性的表达,传递着设计者的主观创造指向与客观环境特性。

　　染织设计是设计教育中最具发展潜质的学科之一,在国家高等学校专业目录修订的大背景下,该学科历久而弥新,始终连接和跨越传统设计与现代设计、材料设计与成品设计、视觉设计与触觉设计、基础教育与专业教育,具有很强的艺术性、实践性。高层次、大范围的设计赛事既促进了学生融合艺术理论与实践技能解决实际问题的能力,又能为他们毕业之后顺利走上社会岗位打下基础、创造平台。"全国纺织品设计大赛暨国际理论研讨会"已成为全国高校染织与服装设计学科发展、展示、交流、实践的平台。

　　发展染织与服装设计学科,培养更多符合时代潮流与社会需要的艺术设计人才,是我们的责任与追求。希望本书的出版能够得到社会更多的关注,促进我国纺织行业设计更快地迈进世界领先行列。

清华大学美术学院院长

2012 年 2 月 22 日

目录
CONTENTS

卷首语　鲁晓波

2012年国际植物染艺术设计展作品 — International Plant Dyeing Art Exhibition Works 2012

- 001　Green Fablic　Ayako Ohmizu
- 002　Earth　Ayako Ohmizu
- 003　HANERI　Aya Karashima
- 004　秋意盎然　崔笑梅
- 005　植物扎染　陈立
- 006　纠结、终极符号　曹敬钢
- 007　Abyss Part IV　Chang, Hee-Kyung
- 008　The Way Spring Reels　Cho, Ju-Eun
- 009　Festival　Cho, Min-Jung
- 010　Singing in the Wind II　Cho, Ye-Ryung
- 011　Moment　Choo, Kyung-Im Joann
- 012　Untitled　Carys Hamer
- 013　浊水流长（源远流长的浊水溪）　陈景林
- 014　花上屋　崔瑶
- 015　自娱自乐　丁敏
- 016　Autumn Gold 2　Dian Widiawati
- 017　In a Harmony　Devi Candraditya Hady / Adinda Hady
- 018　Present of Nature　Euh, Hyun-Ah
- 019　Spinning Straw into Gold　Esti Siti Amanah Gandana
- 020　Blue Borders　Edric Ong
- 021　记忆　龚建军
- 022　彩虹糖的梦　龚雪鸥
- 023　别茶者　黄丽群
- 024　调色板　黄荣华
- 025　中国拴马桩　何飞龙
- 026　Untitled　Heo, In-Yul
- 027　Blossoms are Scattered in the Wind　Hwang, So-Jung
- 028　Trees and Wind　Hyun mi-kyung
- 029　Untitled　Hannah Ricker
- 030　路径　贾玛莉
- 031　Mother's Sunset　Jang, Hae-Sun
- 032　Double Minds　Jang, Jung-Gil
- 033　A Sea Fog（海雾）　Jang Kye Young
- 034　Retrospection　Jeon Dong Won & Park Jung Ley
- 035　UN Armful　Jung, Jee-Hye
- 036　A Wind　Jung, Yun-Sook
- 037　天外天　金媛善
- 038　I Wish…　Kim, Hyun-Jin
- 039　Female Heart　Kim, Jung-Hee
- 040　Jioning and Jioning　Kim, Kyung-Hee
- 041　The Temple　Kim, Mi-Sik
- 042　Forsythia　Kim, Sae Rom
- 043　Tea Time　Kim, Wal-Soon
- 044　Sunset　Kim, Youn-joo
- 045　Harmony　Kahfiati Kahdar
- 046　春天　刘云均
- 047　新月系列冰裂　李晓淳
- 048　雕题黎　李迎军
- 049　生命·巢　刘娜
- 050　水漾　赖美智
- 051　父♀心&情　林青玫
- 052　荷　刘俊卿
- 053　凤凰花开　林幸珍
- 054　Dressed up for You　Lah, Eui-Sook
- 055　Petals of Wind　Lee, Ae-Ja
- 056　Flow VI　Lee, Jae-Kyung
- 057　Vision of Blue　Lee, Jin-Bong
- 058　On the Alameda…　Lee, Jin-Young

059	In the Flower Garden　Lee, Mal-Soon	087	日　朱小珊
060	Baobab　Lee, Min-Jeong	088	生生不息　王懿龙
061	Sound of Nature　Lee, Moon-Hee	089	椇·茶　许韫智
062	上善·水　刘亚	090	昙花　萧静芬
063	空山鸟语　李薇	091	剪衣　徐秋宜
064	飘　马彦霞	092	花·雪　杨颐
065	Pilow Series　Maureen Carr	093	如是观　杨建军
066	Untitled　Michele Wipplinger	094	涓流　于婷婷
067	山水谣　马颖	095	月白·飞白　杨芳
068	晴　毛晨睿	096	Harmoney　Yoo, Ja-Hyung
069	Korean Pojagi　Oh, Myung-Hee Michelle	097	Start of Light　Yoon, Jae-Shim
070	Old Tree　Park, Ha-Na	098	复刻自然　张洁、朱建重
071	Kippy's Garden　Park, Hye-Yeon	099	流年　赵莹
072	衍花·散红　秦寄岗	100	鱼　郑晓红
073	In the Wind　Rhee, Myung-Soog	101	夏日　张树新
074	Indigo Blue Sea… Pear Floral　Ryu, Myung-Sook	102	竹　张宝华
075	原味生活　任晟萱	103	重生　张靖婕
076	南阳汉画《嫦娥奔月》闪秀桂	104	书生　朱微婷
077	天际系列1号日象、天际系列2号月象　沈晓平	105	叠加　朱医乐
078	涟漪　石历丽	106	夕　张红娟
079	BLATHA's Dream 2012　Sohn, Hee-Soon	107	蝴蝶妈妈　杨文斌
080	无题　山崎和树	108	靛蓝染　Kim, Kwang Soo
081	汐　单夏丽	109	Remains　Reiko Hara
082	上善若水　田青	110	母与子　王晶晶
083	水的遐想　吴越齐	111	韵　刘玥
084	消失的记忆　王斌	112	茶染　曹宇坤
085	喜相逢　吴元新	113	Untitled　Sara Ashford
086	暮　吴波	114	阿锦　杨锦雁

民间植物染作品   Folk Plant Dyeing Works

116 拓印 印度（1）	142 植物染色织布 江苏 南通（1）
117 拓印 印度（2）	143 植物染色织布 江苏 南通（2）
118 拓印 印度（3）	144 植物染色织布 江苏 南通（3）
119 拓印 印度（4）	145 植物染色织布 江苏 南通（4）
120 糊染 非洲（1）	146 扎染 靛蓝 云南 大理 白族（1）
121 糊染 非洲（2）	147 扎染 靛蓝 云南 大理 白族（2）
122 香云纱 薯莨染 广州 顺德 汉族	148 扎染 靛蓝 云南 大理 白族（3）
123 靛蓝、薯莨、牛胶染 广西 那坡 黑衣壮族（1）	149 扎染 靛蓝 云南 大理 白族（4）
124 靛蓝、薯莨、牛胶染 广西 那坡 黑衣壮族（2）	150 扎染 靛蓝 云南 大理 白族（5）
125 靛蓝染 广西 那坡 黑衣壮族	151 扎染 靛蓝 云南 大理 白族（6）
126 蓝印花布 靛蓝 "纺织图" 壁挂 江苏 南通 汉族	152 扎染 靛蓝 云南 大理 白族（7）
127 蓝印花布 靛蓝 "飞天" 壁挂 江苏 南通 汉族	153 扎染 靛蓝 云南（1）
128 蓝印花布 靛蓝 "凤戏牡丹" 桌布 江苏 南通 汉族	154 扎染 靛蓝 云南（2）
129 蓝印花布 靛蓝 "年年有余" 工艺品系列 江苏 南通 汉族	155 扎染 靛蓝 云南（3）
130 蓝印花布 靛蓝 江苏 南通 汉族（1）	156 扎染 云南 大理 白族（1）
131 蓝印花布 靛蓝 江苏 南通 汉族（2）	157 扎染 云南 大理 白族（2）
132 蓝印花布 靛蓝 江苏 南通 汉族（3）	158 扎染 云南 大理 白族（3）
133 蓝印花布 靛蓝 江苏 南通 汉族（4）	159 扎染 靛蓝 云南（4）
134 蓝印花布 靛蓝 江苏 南通 汉族（5）	160 扎染 靛蓝 云南（5）
135 蓝印花布 靛蓝 江苏 南通 汉族（6）	161 夹缬 靛蓝 浙江 温州（1）
136 蓝印花布 靛蓝 花卉动物纹 江苏 南通 汉族	162 夹缬 靛蓝 浙江 温州（2）
137 蓝印花布 靛蓝 花卉纹 江苏 南通 汉族	163 夹缬 靛蓝 浙江 温州（3）
138 蓝印花布 靛蓝 几何纹 江苏 南通 汉族（1）	164 夹缬 靛蓝 浙江 温州（4）
139 蓝印花布 靛蓝 几何纹 江苏 南通 汉族（2）	165 夹缬 靛蓝 浙江 温州（5）
140 彩色浇花布 江苏 南通（1）	166 夹缬 靛蓝 浙江 温州（6）
141 彩色浇花布 江苏 南通（2）	167 夹缬 靛蓝 浙江 温州（7）

168	夹缬　靛蓝　浙江　温州（8）	
169	夹缬　靛蓝　浙江　温州（9）	
170	夹缬　靛蓝　浙江　温州（10）	
171	夹缬　靛蓝　浙江　温州（11）	
172	艾德丽丝绸　枸杞根、桑树瘤子、生铁渣染色　新疆　和田	
173	艾德丽丝绸　红柳枝、凤仙花、卡列古丽染色　新疆　和田	
174	艾德丽丝绸　红柳枝、茜草、槐花染色　新疆　和田	
175	艾德丽丝绸　红柳枝、卡列古丽、茜草染色　新疆　和田	
176	艾德丽丝绸　核桃皮、卡列古丽、奥斯曼染色　新疆　和田	
177	蜡染　靛蓝染　广西　苗族	
178	靛蓝染　广西　隆林　苗族	
179	蜡染　靛蓝染　广西	
180	粘膏染　靛蓝染　广西　白裤瑶族	
181	植物染彩色裙（清代）　贵州　从江县　苗族	
182	蜡染刺绣妇女上衣（民国）　贵州　纳雍　苗族	
183	妇女蜡染上衣（民国）　贵州　镇宁　布依族	
184	妇女蜡染刺绣上衣（民国）　贵州　贵定　苗族	

185	蜡染被面（清末）　贵州　三都　苗族	
186	蜡染被面　贵州　丹寨　苗族	
187	蜡染　靛蓝染　贵州　苗族（1）	
188	蜡染　靛蓝染　贵州　苗族（2）	
189	蜡染被面（清代）　贵州　惠水　苗族	
190	衣袖片（民国）　贵州　黔西　苗族	
191	蜡染围腰　贵州　榕江	
192	蜡染刺绣衣袖（民国）　贵州　黄平革家	
193	植物彩色蜡染被带片（民国）　贵州　黔西　苗族	
194	蜡染植物染色拼布儿童包巾（清代）　贵州　惠水　布依族	
195	织金蜡染方巾（现代）　贵州	
196	蜡染祭鼓幡　贵州　榕江县　高排苗族	
197	蜡染　贵州	
198	蜡染　靛蓝、金沟藤、黄栀子等染色　贵州	
199	蜡染　靛蓝染　贵州	
200	蜡染刺绣妇女头帕（民国）　海南　苗族	
201	素纱禅衣（长沙马王堆出土，南京云锦研究所复制）	

2012年国际植物染艺术设计展作品

International Plant Dyeing Art Exhibition Works 2012

- 中国大陆/Mainland China
- 中国台湾/Taiwan, China
- 韩国/Korea
- 日本/Japan
- 美国/America
- 印度/India
- 非洲/Africa
- 印度尼西亚/Indonesia
- 马来西亚/Malaysia

姓名：Ayako Ohmizu
国籍：日本

简历：1980年　出生于东京都
多摩美术大学　Textile Design学科硕士毕业
石垣岛八重山上布　师从新垣幸子
经纬绊绢和服　师从石黑裕子氏
现在以自然治愈为主题进行创作
使用绊（IKAT）技法，进行挂毯、服饰面料、室内装饰的作品的制作。

作品名称：Green Fablic　　材料：真丝、栀子染、桑叶染

ARTIST NAME: Ayako Ohmizu
COUNTRY: Japan

CURRICULUM VITAE:
Hiroko Takano
Graduated from Textile Design Course, Department of Product and Textile Design, Faculty of Art and Design, Tama Art University in 2007.
Graduated from Textile Design, Master's Degree Course, Tama Art University in 2009.
Started working at Textile Design Department as a research assistant of Banana Textile Project, Tama Art University in 2009.

ARTWORK TITLE: Earth   MATERIAL: banana fiber, linen, natural dye   SIZE: 135cm × 135cm

ARTIST NAME: Aya Karashima
COUNTRY: Japan

CURRICULUM VITAE:
After my experiences as a designer at the handloom company, an in-house designer, and a teacher at a junior high school, I have been working as a freelance designer, as well as a part-time lecturer at Tama Art University.
I create daily products and have my works carried at select shops, craft galleries, and so on. The aim of my creation is to convey Japanese traditional culture to the next generation so as to realize functionality for modern living.

ARTWORK TITLE: HANERI   MATERIAL: silk   SIZE: 18cm×100cm

姓名：崔笑梅
国籍：中国

简历：山东工艺美术学院纤维染织专业教师
　　　主要作品有《竹林听风》、《来自音乐的某种快乐》、《网》、《仲夏情怀》等
　　　出版《中国传统图案摹绘精粹》（与人合著）

作品名称：《秋意盎然》　材料：棉布　尺寸：60cm×180cm

姓名：陈立
国籍：中国

简历：1982年毕业于中央工艺美术学院染织美术专业，毕业后留校任教至今。1999年中央工艺美术学院并入清华大学，更名为清华大学美术学院，任该院染织服装艺术设计系副教授、硕士生导师。

作品名称：《植物扎染》　　材料：棉布　　尺寸：160cm×60cm
作品名称：《植物扎染》　　材料：双绉　　尺寸：300cm×100cm

姓名：曹敬钢
国籍：中国

简历：1979年 考入天津工艺美术学校
1986年 考入天津美术学院服装设计专业
1990年 分配到天津外贸丝绸进出口公司从事服装设计工作
1998年 调入天津美术学院服装染织设计系任教至今
2000—2002年 就读于天津工业大学研究生班
本人为中国青海省循化撒拉族自治县政府指定的撒拉族民族服装设计师
设计作品多次入选国内外艺术展
近期参展有2009年日本京都"日中纤维艺术展"
2010年北京"亚洲纤维艺术展"
2010年中、日、韩"亚洲超越展"
2011年北京"世界拼布艺术展"

作品名称：《纠结》　　材料：麻　尺寸：180cm×100cm
作品名称：《终极符号》　材料：麻　尺寸：128cm×110cm

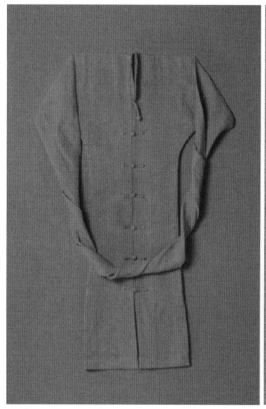
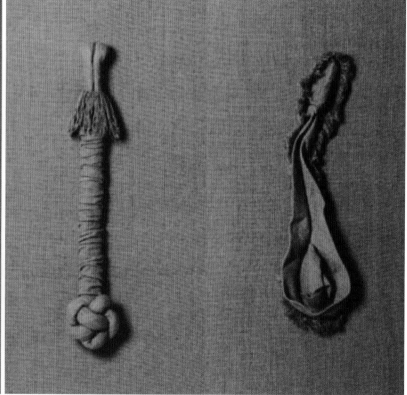

ARTIST NAME: Chang, Hee-Kyung
COUNTRY: Korea

CURRICULUM VITAE:
Director of the Korean Society of Fashion Business
Director, Traditional /Stage Costume & Item <DDAM>
Chief of planning office <MYCO>
2011 International Textiles and Costume Congress <Heritage Textile and Costume>, Institute of Technology, Bandung, Indonesia
2009 International Fashion Exhibition by Invitation <Gloval Fashion & Multi-Culture>, Exhibition Hall, Yi Fu Bulding, Donghua University, Shanghai, China
Present: Lecturer, Department of Clothing & Textiles, Sookmyung Women's University

ARTWORK TITLE: Abyss Part IV   MATERIAL: viscos rayon   SIZE: 50cm × 80cm

ARTIST NAME: Cho,Ju-Eun
COUNTRY: Korea

CURRICULUM VITAE:
EWHA Women's University Industrial design (Ph. D)
SK TELESYS (R & D product design team) senior manager (2003—2007)
New frontier award 2002 (selected - formative art)
The exhibition of design Olympic (product design)(2008)
Good-Design fair 2011 (selected - environment design)
Present: Fiber artist. Metropolitan Architecture Company (Inspace & furniture designer)

ARTWORK TITLE: The Way Spring Reels MATERIAL: silk, sapanwood tree, gardenia, alum SIZE: (35cm×240cm)×2

ARTIST NAME: Cho,Min-Jung
COUNTRY: Korea

CURRICULUM VITAE:
Emily Carr Institute of Art & Design BFA
Ph.D of Clothing Textile, Seoul National University
Li & Fung Korea, Merchandiser
Hansae, Merchandiser & Georgio Ferri, golf wear designer
Present: Fashion Design & Lecturer at Kyungwon University

ARTWORK TITLE: Festival   MATERIAL: Korean linnen (Moshi), sapanwood tree, gardenia, alum

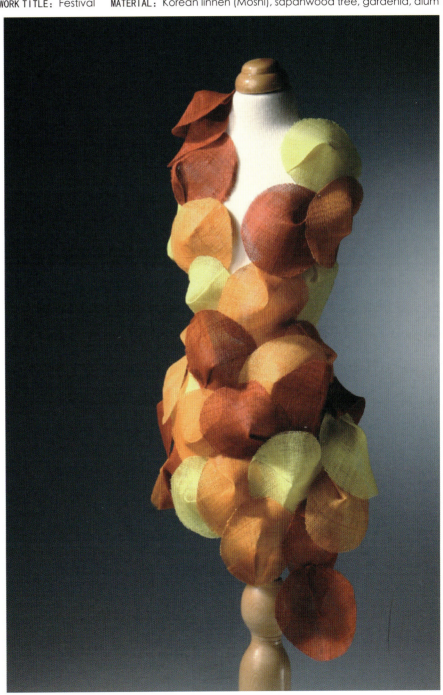

ARTIST NAME: Cho, Ye-Ryung
COUNTRY: Korea

CURRICULUM VITAE:
3 solo exhibitions (Korea)
2010: European Patchwork Meeting 2010 (Hands Of Korea) (Alsace, France)
2011: The 46th Korea Design Exhibition, the award of the Korea Institute of Design Promotion
Present: Lecturer at Dongduk Women's University, Sangmyung University, Shingu University

ARTWORK TITLE: Singing in the Wind II   MATERIAL: Korean paper, silk yarn, tyvek   SIZE: 45cm×45cm

ARTIST NAME: Choo, Kyung-Im Joann
COUNTRY: Korea

CURRICULUM VITAE:
Purdue University, Master of Arts
5 solo exhibitions (America and Korea)
2010: SH Contemporary Art Fair, Shanghai, China
2010: Craft Trend Fair, Seoul, Korea
2009: Korea Galleries Art Fair, Busan, Korea
Present: Instructor at Hanyang Women's University

ARTWORK TITLE: Moment   MATERIAL: hemp cloth, silk, cotton yarn   SIZE: 30cm×30cm

ARTIST NAME: Carys Hamer
COUNTRY: America

ARTWORK TITLE: Untitled   MATERIAL: silk scarf   SIZE: 19cm×160cm

姓名：陈景林
国籍：中国

简历：台湾师范大学美术研究所西画组硕士
天染工坊共同创办人及艺术总监
曾获第六届南瀛奖首奖，第一届、第三届民族工艺奖三等奖、二等奖。
《大地之华台湾天然染色事典》一、二册，《染织编绣巧天工》、《植物炼金术》等。
主要研究天然染色、纤维材质学、少数民族服饰及染织绣工艺、纤维艺术。

作品名称：浊水流长(源远流长的浊水溪)　　材料：亚麻布、蓝靛染料 (linen, indigo)　　尺寸：375cm × 200cm

姓名：崔瑶
国籍：中国

简历：来自黑龙江省。2008年进入清华大学美术学院染织服装系学习。现为染织专业大四学生。
作品曾获"2011全国纺织品设计大赛暨国际理论研讨会"铜奖及优秀奖、"海门全国纺织品大赛"优秀奖。

作品名称：《花上屋》　　材料：棉　苏木、黄连　　尺寸：50cm×50cm

姓名：丁敏
国籍：中国

简历：广州美术学院服装设计专业硕士
长年从事染织艺术设计教学工作，现任广州美术学院工业设计学院纤维艺术设计工作室负责人。
致力于基于地域文化的传统手工技艺的创新设计研究。设计作品历年参加国内外专业展览、大赛、设计周等，屡获奖项。

作品名称：《自娱自乐》　　材料：棉　　尺寸：73cm×30cm×75cm

ARTIST NAME: Dian Widiawati
COUNTRY: Indonesia

CURRICULUM VITAE:

Institut Teknologi Bandung (ITB), Faculty of Art & Design, Textile Design Studio (Graduated)
Master Degree, Design Departement, Institute of Technology, Bandung (ITB)

ARTWORK TITLE: Autumn Gold 2 **MATERIAL:** silk and natural dyeing sappan wood (Caesalpinia sappan LINN), Tegeran (Cundraina javanensis), gambier (Uncaria gambier, Roxb) **SIZE:** 70cm×220cm

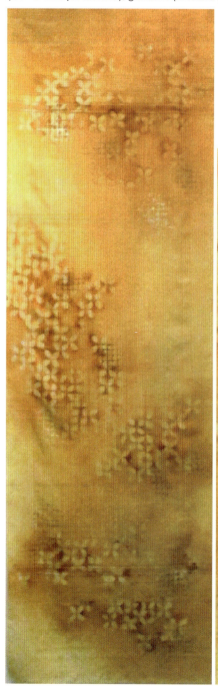
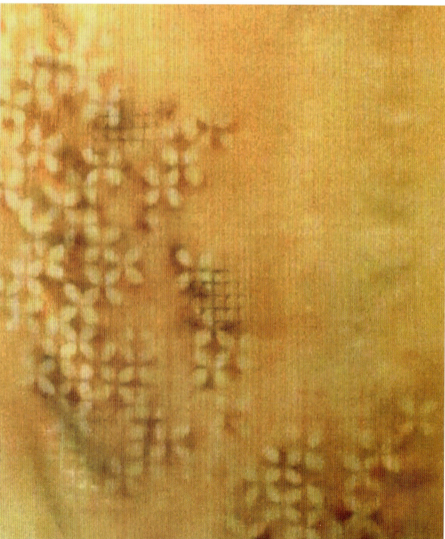

ARTIST NAME: Devi Candraditya Hady / Adinda Hady
COUNTRY: Indonesia

CURRICULUM VITAE:
I am currently doing a bachelor in Textile Arts and Design, Faculty of Arts and Design, Bandung Institute of Technology, Indonesia. I have been through many experiences in my study and some projects. My recent project for my intership is making a collection of Fashion Show Trend 2011/2012 for one of Indonesian Designer, using ribbon threads and macrame as a technique. According to that, I made more research for my study about exploration of natural plant dyeing (soga tingi) on China grass (rami) using macrame as a technique.

ARTWORK TITLE: In a Harmony [Ceriopstagal (Perr.) C. B. Rob.]
MATERIAL: natural fiber; China grass (rami), natural plant dyeing; Soga Tingi
SIZE: (25cm×150cm) ×2

ARTIST NAME: Euh, Hyun-Ah
COUNTRY: Korea

CURRICULUM VITAE:
2011 PH.D Sangmyung University
2002—2009 2nd Solo Exhibitions
2002—2003 21th, 22th Korea Craft Competition Special Selection
1996—2012 Group & Invitational Exhibitions
Lecturer, Department of Design Sangmyung University
Director of Design Unique One

ARTWORK TITLE: Present of Nature　　MATERIAL: straw, hemp cloth, wool, biz / grape dyeing, pear dyeing　　SIZE: 40cm×40cm

ARTIST NAME: Esti Siti Amanah Gandana
COUNTRY: Indonesia

CURRICULUM VITAE:
Assistant Lecturer at Bandung Institute of Technology (ITB), Faculty of Art and Design, Textiles Craft, Bandung-Indonesia.
Consultant: Fashion, Textiles, Craft [Natural Fibres, Dyes, and Textile Stucture Design (Knitting)]

ARTWORK TITLE: Spinning Straw into Gold   MATERIAL: 100% wheat straw   SIZE: 30cm×120cm

ARTIST NAME: Edric Ong
COUNTRY: Malaysia

CURRICULUM VITAE:
Aid to Artisans Advocate Award 2006
Seals of Excellence for Craft Products from UNESCO-AHPADA 2001—2007
Japanese G-Mark for Good Design in Lifestyle/Home Accessories
Penyokong Kraf Negara (Malaysian National Non Governmental Organization Award)2007
MALAYSIAN DESIGNER OF THE YEAR AWARD 2009 (MERCEDES BENZ/STYLO)
STYLO Kuala Lumpur Fashion Festival Heritage Award 2008
Australia Culture Award
Pegawai Bintang Sarawak (Officer of the Star of Sarawak),Malaysia Award from the Sarawak State Government
President of Society Atelier Sarawak, Malaysia (since 1999)
Vice-President, World Crafts Council Asia Pacific 2008
Board Member, Crafts Council of Malaysia
Immediate Past President of the ASEAN Handicraft Promotion and Development Association (AHPADA)
Consultant to UNESCO on Heritage, Architectural Conservation and Crafts
Jury member of UNESCO-AHPADA Seal of Excellence for Crafts
Curator of international exhibitions on crafts and textiles Architect [Edric Ong Architect;
Malaysian Association Architects(Sarawak); Kumpulan Filipino Malaysian Architect]
Author and International speaker on Architectural Conservation, Crafts and Textiles
Founder and Convener of World Eco-Fiber and Textile (WEFT) Forum 1999, 2001 and 2003 and 2008 held in Kuching Sarawak Malaysia

ARTWORK TITLE: Blue Borders   MATERIAL: silk, natural indigo dye (marsdenia tinctoria)   SIZE: 700mm×2000mm

姓名：龚建军
国籍：中国

简历：1954年生于江苏南通，高级工艺美术师，江苏省工艺美术大师，南通三友民间艺术研究所所长，原在南通工艺美术研究所从事传统手工印染技艺方面的研发20多年。现从事残疾人技能培训工作。

作品名称：《记忆》　　材料：全棉灯芯绒布　　尺寸：29cm×46cm

姓名：龚雪鸥
国籍：中国

简历：2004年参加第三届"'从洛桑到北京'——国际纤维艺术双年展"
2005年获"中国国际家用纺织品设计大赛暨国际理论研讨会"创意设计优秀奖
2009年、2010年、2011年获第九届、第十届、第十一届"全国纺织品设计大赛暨理论研讨会"论文优秀奖

作品名称：《彩虹糖的梦》　　材料：羊毛　　尺寸：260cm×50cm

姓名：黄丽群
国籍：中国

简历：就读于广州美术学院染织艺术设计纤维工作室。2012年本科应届毕业生。热爱传统工艺，尊重自然的生命力，追求古今结合的创意设计、返璞归真的生活态度。

作品名称：《别茶者》　　材料：棉麻布　　尺寸：2m×3m

姓名：黄荣华
国籍：中国

简历：男，55岁。大学本科，工程师。现为常州云卿纺织品公司总工，北京菲怡芳馨公司技术总监。拥有自己的天然染色工作室。1993年开始进行植物染料染色研发工作，发表天然染色专业文章数百篇。多家媒体对其有采访。

作品名称：《调色板》　　材料：丝绸　　尺寸：50cm×35cm

姓名：何飞龙
国籍：中国

简历：2007—2011年 就读于西安美术学院服装系（本科）
　　　2011年 考入西安美术学院服装系攻读硕士研究生

作品名称：《中国拴马桩》　材料：综合材料　尺寸：(50cm×250cm)×2

ARTIST NAME: Heo, In-Yul
COUNTRY: Korea

CURRICULUM VITAE:
2003 M.A. Graduated from School of Art & Design Sangmyung University Korea
2006 Professional Development Diploma-Tapestry weaving
West Dean College England
Solo Exhibition twice / Many Group Exhibition at home and abroad
Web exhibition:
http://www.americantapestryalliance.org/Exhibitions/Inyulheo/Welcome.html

ARTWORK TITLE: Untitled   MATERIAL: Korean paper, silk   SIZE: 40cm×40cm

ARTIST NAME: Hwang, So-Jung
COUNTRY: Korea

CURRICULUM VITAE:
M.A. Graduated from School of Art & Design Sangmyung University
2011 Solo Exhibition (Seoul, Korea)
     The 17th Sangmyung Designer Exhibition (Seoul, Korea)
2010 The 16th Sangmyung Designer Exhibition (Seoul, Korea)
     JAPANTEX (Tokyo, Japan)
2009 The 15th Sangmyung Designer Exhibition (Seoul, Korea)

ARTWORK TITLE: Blossoms are Scattered in the Wind   MATERIAL: HanJi, cotton yarn   SIZE: 100cm×40cm

ARTIST NAME: Hyun mi-kyung
COUNTRY: Korea

CURRICULUM VITAE:
BFA in Art Education, Jeju National University, Jeju, Korea
Currently Recommended Artist of National Art Exhibition Jeju Special Self-Governing Province
Member of the Jeju Art Association, Jeju
Member of the Jeju Textile Artists Association, Jeju
Member of C.Q.A (Corea Quilt Association)
Exhibition of Internation Quilt Festival, Seoul, South Korea (2nd to 4th, 6th)
2011.8 "Creation and Parody" Exhibition in Jeju Museum of Art
2010.2 Hyun mi-kyung's Solo Exhibition "Story of Needlework"

ARTWORK TITLE: Trees and Wind   MATERIAL: fabric dyed with water of persimmons   SIZE: 147cm×27cm

ARTIST NAME: Hannah Ricker
COUNTRY: America

CURRICULUM VITAE:
I have always been drawn to color, texture and patterns. As a child, I remember spending countless hours looking up at the leaves in a tree, endless pebbles on a beach, and undulating jellyfish at the aquarium. The organic and amorphous lines, shapes and patterns in nature continue to inspire me. Surface design and relief printing are my favorite mediums to work in. They are completely different, yet have many similar and overlapping qualities. I am happy to have discovered the beauty and immense color of natural dyes, and look forward to continue exploring with it in my art.

ARTWORK TITLE: Untitled   MATERIAL: tussah silk   SIZE: 50cm×140cm

姓名：贾玛莉
国籍：中国

简历：2007年12月　台中县立文化中心"台湾染织协会纤维艺术展"蓝靛梭织餐桌垫组
　　　2008年12月　台中县立文化中心"台湾染织协会纤维艺术展"双重织色彩游
　　　2008年4月　彰化县文化局"纤语——纤维艺术展"双重织色彩游
　　　2009年8月　台中市文英馆"纤语——纤维艺术展"双重织rosepath 路径
　　　2011年4月　"法国 La Rochelle ISEND 2011天然染色织品双年展"

作品名称：《路径》　　材料：日本和纸、日本苎麻 姜黄、紫胶虫、七里香、洋葱皮、茜草、蓝草　　尺寸：75cm×180cm

ARTIST NAME: Jang, Hae-Sun
COUNTRY: Korea

CURRICULUM VITAE:
HanYang University social education center dyeing fibre art finished
KyungBok Palace (景福宮) traditional dress and its ornaments reproduction process finished
The Exhibition of Gallery Blue "Gyubang Craft" (seoul)
The Invitation of Exhibition at Dobu Department store in Tokyo(2002)
International Natural-dyeing Invitation of Exhibition (Ulsan, DaeGu)(2007)
Natural-dyeing exhibition "An Outing" 2011(Seoul Arts Center Artshop Gallery)
Present: Natural-dyeing & Korean traditional patch-work artist

ARTWORK TITLE: Mother's Sunset SIZE: 82cm×97cm
MATERIAL: Korean silk organza (Ock-sa), fixatives(alum, ferrous sulfate)

ARTIST NAME: Jang, Jung-Gil
COUNTRY: Korea

CURRICULUM VITAE:
Graduated from OTSUKA Textile Design Institute (1994)
Awarded prize at "THE JAPAN TEXTILE CONTEST"
Korean Textile Association Exhibition (1994—2008)
International Gift Show Exhibition 2002 (Japan, Tokyo)
International Exhibition of Natural dyed work For Invitational & Exchange (2008)
Present: Art director, fiber artist

ARTWORK TITLE: Double Minds   MATERIAL: silk, lac, ferrous sulfate   SIZE: 115cm×280cm

ARTIST NAME: Jang Kye Young
COUNTRY: Korea

CURRICULUM VITAE:
Worked as a staff of Broadcasting center Hanwha63city (1985—1990)
Majored in Economics at Korea National Open University
Completed natural dyeing at Gangwon Polytechnic Institute (2010)
Completed special dyeing high level course at Leepo Dyeing Institute(2011)
Natural dyeing exhibition "An Outing" 2011(Artshop Gallery of Seoul Arts Center)
Present: Natual-dyeing artist (coffee-dyeing)

ARTWORK TITLE: A Sea Fog (海雾)　　MATERIAL: Korean silk (Myung-Ju)
SIZE: (35cm×360cm) ×2

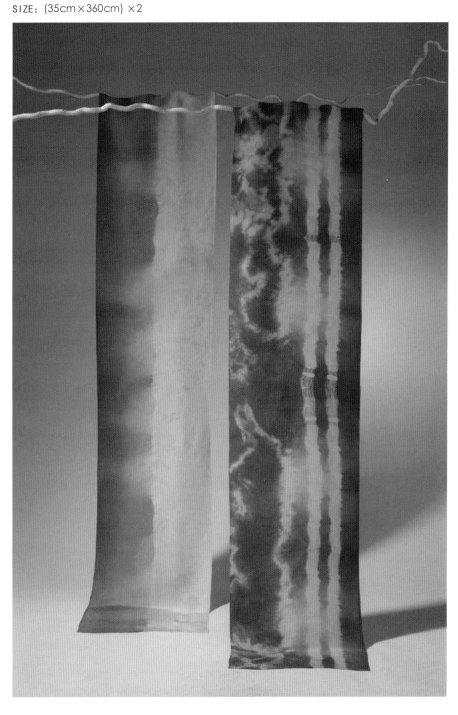

ARTIST NAME: Jeon Dong Won
COUNTRY: Korea

CURRICULUM VITAE:
majored in polymer chemistry at university and graduate school. Since 1983, he has been working in the Ewha Womans University as a professor. Current researches are centered on the satisfaction of the health care in the textiles, and on the natural dyeing.

ARTIST NAME: Park Jung Ley
COUNTRY: Korea

CURRICULUM VITAE:
majored in Fiber Art at Ewha Womans University. Since 1994, she has been working in the Hanyang Womans University as a professor (Textile Fashion Design). Solo exhibition 6 times.

ARTWORK TITLE: Retrospection
MATERIAL: natural dyeing material (Caesalpinia sappan, indigo, gromwell), silk fabric (organza, habutae)   SIZE: 170cm × 100 cm

ARTIST NAME: Jung, Jee-Hye
COUNTRY: Korea

CURRICULUM VITAE:
Graduated required full course from dress designing subject at MJM Dress Designing Academy in France
(Specialized in Stylisme / Mdoelisme part)
Study at clothing subject, the postgraduate school of Sukmyung Womens University

ARTWORK TITLE: UN Armful   MATERIAL: silk(80%) rayon(20%)   SIZE: 130cm × 40cm

ARTIST NAME: Jung, Yun-Sook
COUNTRY: Korea

CURRICULUM VITAE:
Korean Traditional Costume Exhibition (Washington)
Korea, Mongolia traditional Costume Festival
Korean traditional curtain "Bal" exhibition (Japan, gallery LOZYE) 2007
Korean traditional curtain "Bal & Bozagi" exhibition (Japan, Hirosima/ The old Bank of Japan)2001
Domestic and abroad a lot of group exhibition and winning a prize International
Present: Korean traditional artshop & center "Chaedamjung" Representation
Korean traditional patch-work artist

ARTWORK TITLE: A Wind   MATERIAL: Korean silk (Myung-Ju & Ock-sa), Korean linnen (Moshi)   SIZE: 134cm×215cm

姓名：金媛善
国籍：中国

简历：被誉为"中国拼布第一人"的朝鲜族拼布艺术家金媛善女士(64岁)自幼受外祖母、母亲的熏陶，喜爱手工制作。20多年来一直从事中国朝鲜族拼布艺术的研究和创作。多次参加日本、韩国和美国的拼布艺术展览，并获国际拼布二等奖。2006年、2009年两次应邀在韩国举办个人作品展。作品先后被韩国、日本、美国等艺术机构和个人收藏。两次在清华大学美术学院演讲，并应邀于2009年3月在清华大学、2010年10月在北京服装学院举办个人作品展。金媛善女士的拼布作品犹如春风拂面，让国人体会到中国文化独特的魅力与韵味。"金媛善女士是目前唯一能代表中国手工拼布艺术水平的艺术家"——这是中国工艺美术学会唐克美副理事长观看展览后的第一感受。中国工艺美术博物馆吕品田馆长在留言簿中写道"巧夺天工"。

先后被韩国江陵市政府聘请为名誉顾问，被北京服装学院服装博物馆聘请为专家顾问，担任亚洲妇女交流协会理事。

作品名称:《天外天》　　尺寸：180cm×200cm

ARTIST NAME: Kim, Hyun-Jin
COUNTRY: Korea

CURRICULUM VITAE:
B. F. A.,Department of Fine Arts SungKyunKwang University (1982)
University art exhibition special prize award(1981)
Leepo Studio of textile (Basic, Special, Natural) Dyeing process finished (2005—2008)
Coffee Dyeing Workshop (Inchon Bugwang Girl's High School)
Natural-dyeing exhibition "An Outing" 2011 (Seoul Arts Center Artshop Gallery)
Present: Fiber artist. "Jei" a publishing company Representative

ARTWORK TITLE: I Wish…   MATERIAL: silk, lac, gardenia, Indigo blue, ferrous sulfate   SIZE: 110cm×110cm

ARTIST NAME: Kim, Jung-Hee
COUNTRY: Korea

CURRICULUM VITAE:
Graduated from ChangWon-MoonSung College of Fashion Department
KYONNAM and HAPUCHON Art Village Opening Commemoration Invitation Exhibition (2005)
YoungNam Part of Social Education Center Fashion Painting Department Chairman (2006)
France (Paris) Invitation Author Exhibition of Academy of Painting Pictures and Writings (2010)
The 3rd Northern & Southern Unification Time of the World Art Contest winning a grand prize (2011)
Domestic and abroad a lot of group exhibition and winning a prize International
Present: Fiber artist. Arts and Crafts "Jung-Whi" Research center Representation

ARTWORK TITLE: Female Heart   MATERIAL: persimmon, Indigo blue, dyeing without toxic lacquer

ARTIST NAME: Kim, Kyung-Hee
COUNTRY: Korea

CURRICULUM VITAE:
Ph. D., Department of Textile Design, Hanyang University
Rabiz Fabric Interior Design Planning section chief
Annyang Science university Department of Fashion Design Invitation professor (2004—2006)
The 2th Kyung Hee, Kim Private Exhibition
Natural dyeing exhibition "An Outing" 2011 (Artshop Gallery of Seoul Arts Center)
Domestic and abroad a lot of group exhibition and winning a prize International
Present: Fiber artist. Lecturer at Hanyang University

ARTWORK TITLE: Jioning and Jioning   MATERIAL: Korean silk organza (Ock-sa), sapanwood tree, gardenia, alum   SIZE: 125cm×170cm

ARTIST NAME: Kim, Mi-Sik
COUNTRY: Korea

CURRICULUM VITAE:
9 solo exhibitions (Korea, America, Japan, China)
2011 International Quilt Festival 2011 "Master of Artquilt" Huston, America
2011 International Quilt Exhibition & Conference Beijing, China
2010 European Patchwork Meeting 2010 (Hands of Korea), Alsace, France
2009 "New wind from Asia" Quilt national Museum, Paducah, America
Present: Profilts Coessor, Sookmyung Women's University Chung Young Yang Embroidery Museum Learning Center
Oversea Expert of China Qulor and Art Committee
Representative of SAQA (Studio Art Quilt Associates)

ARTWORK TITLE: The Temple   MATERIAL: Chinese ink dyed cotton   SIZE: 170cm×190cm

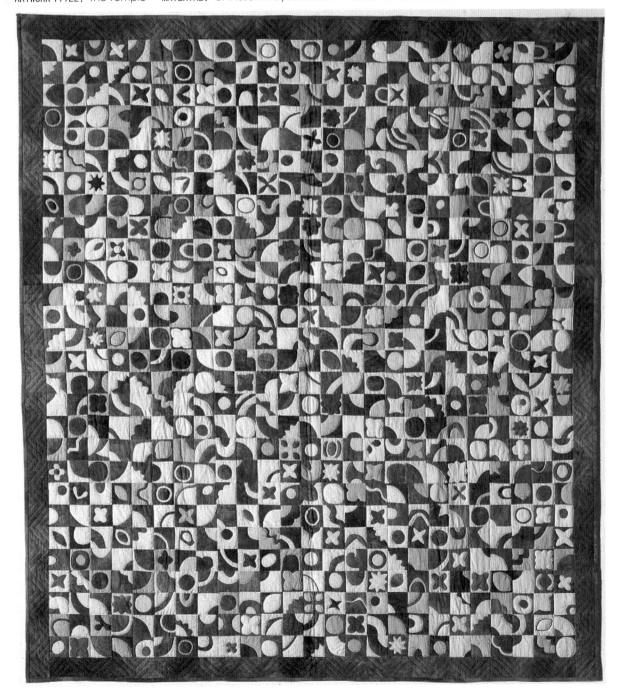

ARTIST NAME: Kim, Sae Rom
COUNTRY: Korea

CURRICULUM VITAE:
Master degree in Sookmyung MYUNG Women's University

ARTWORK TITLE: Forsythia   **MATERIAL:** silk   **SIZE:** 80cm×30cm

ARTIST NAME: Kim, Wal-Soon
COUNTRY: Korea

CURRICULUM VITAE:
Ph.D. Department of Clothing, Sungshin Women's University
The 5th Wolson, Kim Exhibition & Fiber Modeling Jackpot Exhibition gold prize(2006)
Domestic and abroad a lot of group exhibition and winning a prize International
Special Selection Prize at The Korea Craft Exhibition awards
1993—2001 Korean Fine Arts Association Exhibition & The Korea Craft Council Exhibition
Natural-dyeing exhibition "An Outing" 2011 (Artshop gallery of Seoul Arts Center)
Present : Fiber artist & professor, Department of Fashion Design, Suwon Women's College

ARTWORK TITLE: Tea Time   MATERIAL: silk, Korean linen, alginate hoil, gallnut, "Tinta China", alum, ferrous sulfate   SIZE: 72cm×72cm

ARTIST NAME: Kim, Youn-joo
COUNTRY: Korea

CURRICULUM VITAE:
2012 Completion graduated course at Sookmyoung Women's University
2010 Participation in the International Fashion Exhibition, The costume Culture Association, Korea
2011 Participation in the International Textile and Fashion Costume Exhibition, Bandung, Indonesia

ARTWORK TITLE: Sunset   MATERIAL: silk

ARTIST NAME: Kahfiati Kahdar
COUNTRY: Indonesia

CURRICULUM VITAE:
International Exhibition ENGLAND-KOREA-INDONESIA Art Textile 2011 Bandung
Fiber Face, Yogyakarta 2011
The 4th Surin International Folklore Festival in Surindara Rajabhtat University, Surin Province, Thailand 2009
Asian Fiber Art Axhibition VI, Jakarta, Indonesia 2008
Asian Fiber Art Axhibition V, Okonawa, Japan 2007
Fiber Art Contemporary Exhibition 2005
 "Dua Generasi"
4th 25 June, Gracia Surabaya
Posting Fiber Exhibition 2005
12-21 January, Bentara Budaya Jakarta
20-20 February, Galery Kita, Bandung

ARTWORK TITLE: Harmony

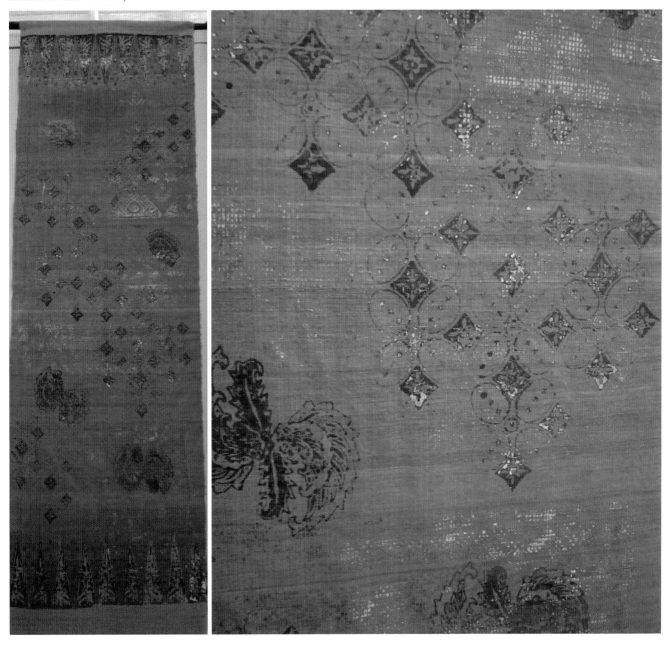

姓名：刘云均
国籍：中国

简历：工程师，1940年出生，现任江苏南通三友民间艺术研究所技术总监，长期从事物理、染化、丝网印工艺的研发工作。

作品名称：《春天》　　材料：真丝素缎　　尺寸：40cm×45cm

姓名：李晓淳
国籍：中国

简历：鲁迅美术学院染织服装艺术设计系硕士研究生
作品入选"2011年国际拼布展"（北京）、"第八届亚洲纤维展"（马来西亚）

作品名称：《新月系列冰裂》　　**材料：**真丝、麻、绢、珠子　　**尺寸：**164cm×117cm

姓名：李迎军
国籍：中国

简历：服装设计学硕士，清华大学美术学院副教授。致力于"民族文化与服装设计"的研究，设计作品《绿林英雄》、《线路地图》、《精武门》、《满江红》、《美人计》、《霸王别姬》荣获多项国际、国内专业设计比赛金、银及国家奖。

作品名称：《雕题黎》　　材料：黎族植物缬染裙、棉布　　尺寸：180cm×60cm×60cm

姓名：刘娜
国籍：中国

简历：纤维艺术作品曾4次参加"'从洛桑到北京'——国际纤维艺术双年展"，获得铜奖一次，获得优秀奖两次。拼布艺术作品参加2011年"'传承与创新'国际拼布艺术展"。多篇纤维艺术方面相关论文发表于重要刊物和论文集中。

作品名称：《生命·巢》　　材料：棉布　　尺寸：50cm×40cm×20cm

姓名：赖美智
国籍：中国

简历：出生于中国台湾中部，现为亚洲大学研究所学生。
从事蓝染工艺，且为台湾蓝四季研究会成员，将天然染色的布料运用在时装创作是其兴趣与喜爱。

作品名称：《水漾》　　材料：100%麻

姓名：林青玫
国籍：中国

简历：1998—2002年 旅日十多年 专门研究服饰与造型设计
现任台湾亚洲大学时尚设计学系系主任
日本东京文化女子大学大学院被服环境学博士、被服学硕士、服装学学士
2009年至今 台湾编译馆专科及职业学校教科书审查委员
2008—2010年 台湾岭东科技大学流行设计系／所副教授兼系主任、所长
2008年至今 国际学术期刊 Textile Research Journal, America (SCI) 审查委员
2007年至今 台湾中华美学教育学会常务理事
2006年至今 台湾工艺研究发展中心、台湾美术馆、台湾"经济部工业局"等评审委员
2005—2007年 台湾东森得易购股份有限公司纺品部顾问
2003年至今 北京电影学院动画学院客座教授
2003年至今 台湾女装、美容、美发技能竞赛裁判

作品名称：《父♀心＆情》　　材料：棉布　　尺寸：170cm

姓名：刘俊卿
国籍：中国

简历：1965年生于中国台湾
亚洲大学数位媒体设计学系研究所硕士班时尚设计组一年级研究生
卓也小屋手工坊设计师
植物染、蓝染教学、产品设计及创作

作品名称：《荷》　　材料：麻布　　尺寸：80cm×118cm

姓名：林幸珍
国籍：中国

简历：她热爱纤维创作，灵感来自大自然，投入教学研究及染织设计，主持社团和独立策展计画等专业工作。连接创作每个细节必须贯穿，追求极致的精神，不断地让艺术生涯汲取知识，释放生命能量，实现理想。获得与人分享的快乐。

作品名称：《凤凰花开》　　材料：植物染布、天然棉衬　　尺寸：145cm×167cm

ARTIST NAME: Lah, Eui-Sook
COUNTRY: Korea

CURRICULUM VITAE:
Former Associate professor, Department of Design, Daewon Science College
Concurrent Professor, Kyung Hee University
KTTE Committee Member

ARTWORK TITLE: Dressed up for You   MATERIAL: silk, plastic, paper   SIZE: 40cm×80cm

ARTIST NAME: Lee, Ae-Ja
COUNTRY: Korea

CURRICULUM VITAE:
1977 B.F.A Department of Applied Arts, Seoul National University
1988 M.F.A Graduated from School of Arts, Daegu Catholic University
1992 California College of Arts (Oakland, America)
10 solo exhibitions (1994—2011)
More than 200 group exhibitions (1988—2011)
Present: Professor, Department of Textile Design, Gyeongnam National University of Science and Technology, Korea

ARTWORK TITLE: Petals of Wind   MATERIAL: silk   SIZE: 50cm×180cm

ARTIST NAME: Lee, Jae-Kyung
COUNTRY: Korea

CURRICULUM VITAE:
2 solo exhibitions
The 2nd-5th from Lausanne to Beijing International Fiber Art Biennale
Present: Ph. D candidate Hongik University

ARTWORK TITLE: Flow VI   MATERIAL: wool   SIZE: 42cm×53cm

ARTIST NAME: Lee, Jin-Bong
COUNTRY: Korea

CURRICULUM VITAE:
Graduated from Otsuka Textile Design Institute (1994)
The 4th Private Exhibition & a lot of group show and winning a prize background
Art fabric performance five artists at "On Air West" 1994(Japan/ Tokyo)
International Natural-dyeing Invitation Exhibition (Ulsan/2007)
Natural dyeing exhibition "An Outing" 2011 (Artshop Gallery of Seoul Arts Center)
Mr. Jung, Myung-Hoon direction opera "IDOMENEO" (Mozart) leading role costume dyeing and appearance costume painting or taking pictures of all the members
Present: Art derector. Fabre artist. Lecturer at Chung-Ang University
Weaving & Dyeing Studio "Leepo" research Representation

ARTWORK TITLE: Vision of Blue
MATERIAL: Korean linnen (Moshi), Indigo blue   SIZE: 105cm×177cm

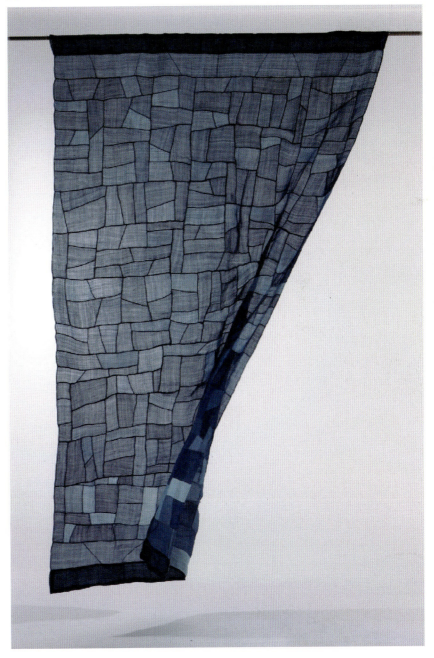

ARTIST NAME: Lee, Jin-Young
COUNTRY: Korea

CURRICULUM VITAE:
B.F.A. Department of Drama and Movie Hanyang University
Leepo Studio of Terxtile (Basic, Special, Natural) Dyeing process finished
Part of Fashion & Interior process finished Inchon Citizen academy
Sansung group leading person training Natural-dyeing workshop (Namisum)
Natural-dyeing exhibition "An Outing" 2011 (Seoul Arts Center Artshop Gallery)
Present: Fabre artist, Junior high school lecturer (Inchoen)

ARTWORK TITLE: On the Alameda...   MATERIAL: silk, indigo blue, lac   SIZE: (35cm×35cm)×2

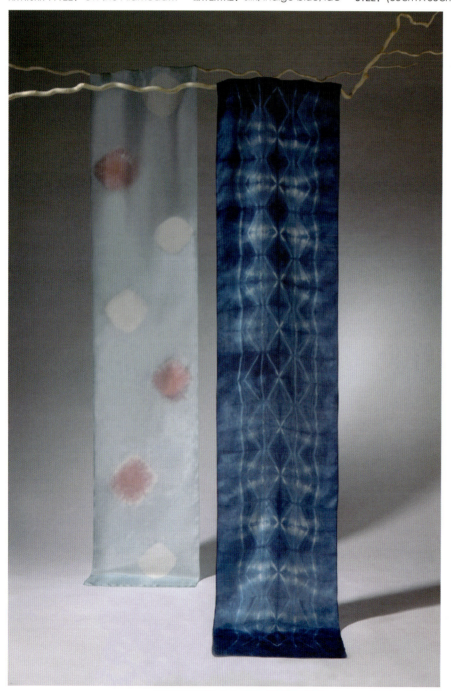

ARTIST NAME: Lee, Mal-Soon
COUNTRY: Korea

CURRICULUM VITAE:
(Association) Korean Imperial Court dress and its ornaments boffin finished
ShI-Ra University Traditional Dyeing Academy specialist course taking (2003—2005)
Natural dyeing leader qualification certificate taking (2006)
The 11th Korean Imperial Court dress and its ornaments boffin exhibition
Korean traditional culture art for 2009 years and winning in 2011
Present: Natural dyeing artist. (Association) Korean Traditional Culture of Art Union female author member

ARTWORK TITLE: In the Flower Garden   MATERIAL: Korean silk (Ock-sa), sapanwood tree, sardenia, lac, Indigoblue, etc.
SIZE: 130cm×180cm

ARTIST NAME: Lee, Min-Jeong
COUNTRY: Korea

CURRICULUM VITAE:
2011 Heritage Textiles and Costume (Indonesia)
The 16th International Invited Fashion Exhibition of Professor (Taiwan, China)
Digital Fashion Exhibition (Osaka)
Present: Lecturer, Sookmuyung Women's University

ARTWORK TITLE: Baobab   MATERIAL: silk, linen, cable   SIZE: 55cm×200cm

ARTIST NAME: Lee, Moon-Hee
COUNTRY: Korea

CURRICULUM VITAE:
Graduated from Ewha Women's University (Major, Fabric Art) (1991)
Graduated first in ESMODE PARIS(2007)
Win a Prize at NIKE's Public Competition for Corset(2006)
Participate in Premier Vision, "Forum Expofil Tendance" (Paris) (2007)
Opening Fashion Show at the 8th Seoul International Movie Festival(2008)
Domestic and abroad lot of group show and winning a prize International
Present: Art director. Fashion designer. Director of the Association of Korean Knitting Design

ARTWORK TITLE: Sound of Nature   MATERIAL: sheepskin, silk, sapanwood tree, alum   SIZE: free size

姓名：刘亚
国籍：中国

简历：就读于清华大学美术学院染织服装系染织方向，曾获"第十一届全国纺织品设计大赛"铜奖及优秀奖。

作品名称：《上善·水》　　材料：丝绸　　尺寸：9m×1.2m

姓名：李薇
国籍：中国

简历：现任清华大学美术学院教授、硕士生导师，留法访问学者
　　　2001年《清、远、静》壁挂获"'从洛桑到北京'——第六届国际纤维艺术双年展"金奖
　　　2004年《夜与昼》服装获"第十届全国美术展览"金奖
　　　1998年《大漠孤烟》服装获第二届"新西兰羊毛杯"铜奖
　　　1998年《流波曲》服装获中国国际服装节"大连杯"银奖

作品名称：《空山鸟语》　　材料：真丝绡、水纱　　尺寸：300cm×120cm

姓名：马彦霞
国籍：中国

简历：天津美术学院服装染织系副教授
中国美术家协会天津分会会员
中国工艺美术学会会员
2009年 作品《延续》入选"第十一届天津市美展"
2009年 作品《蛹1-4》参加"亚洲联盟超越设计展"（展出地点：中国大陆、中国台湾，韩国，日本）
2010年 作品《红与黑》入选"第七届亚洲纤维艺术作品展"

作品名称：《飘》　　材料：棉布、植物染色　　尺寸：80cm×80cm

ARTIST NAME: Maureen Carr
COUNTRY: America

CURRICULUM VITAE:
I have worked with natural dyes for over ten years. During that time, I studied with a group of artists who were mentored by Michele Wipplinger. My art exploration includes painting and printing exclusively using sustainable plant-based dye extracts. I particularly enjoy the unique rich quality of this eco-color palette. The primary focus of my art is home interiors. My favorite colors are madder red and indigo blue. I work in stitched resist and other shibori techniques embellishing with stitching, beads, buttons and shells.

ARTWORK TITLE: Pilow Series   MATERIAL: silk noil   SIZE: 25cm×34cm 33cm×40cm

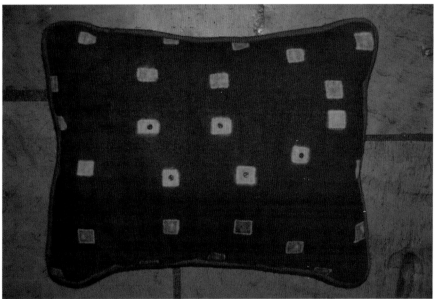

ARTIST NAME: Michele Wipplinger
COUNTRY: America

CURRICULUM VITAE:
Michele Wipplinger is the owner of Earthues, a fair-trade, woman-owned business, working in partnership with artisans and industry for thirty years. The aim of Earthues is to teach and learn about natural dyes and sustainability using eco-methods and exquisite colors for creating beautiful surface designs. Michele provides expertise in color, textile design and artisan craft development for the global marketplace using natural dyes, and assists visionary companies in developing environmentally sound methods for creating and coloring textiles.

ARTWORK TITLE: Untitled   MATERIAL: silk gauze shawl   SIZE: 100cm×170cm   ARTWORK TITLE: Untitled   MATERIAL: silk scarf   SIZE: 21cm×122cm

ARTWORK TITLE: Untitled   MATERIAL: cotton   SIZE: 20cm×35cm   ARTWORK TITLE: Untitled   MATERIAL: silk noil   SIZE: 25cm×26cm

姓名：马颖
国籍：中国

简历：清华大学美术学院2008届本科生
　　　　染织服装艺术设计专业
　　　　清华大学美术学院2012届免试推荐研究生
　　　　作品《悠然畅想》被评为"2011全国纺织品大赛"银奖
　　　　清华大学学生会优秀干部

作品名称：《山水谣》　　**材料**：绡、蓝靛　　**尺寸**：1.8m×2.2m

姓名：毛晨睿
国籍：中国

简历：2008年考入清华大学美术学院染织服装设计系，本科在读。

作品名称：《晴》　材料：棉麻、粗麻　尺寸：1.2m×4m

ARTIST NAME: Oh, Myung-Hee MIchelle
COUNTRY: Korea

CURRICULUM VITAE:
15 solo exhibitions (Korea, China, Japan, America, Spain)
Shanghai Art Fair, Art Shanghai, SH contemporary (China)
Holland Paper Biennale (Neterland)
New York Art Expo (America)
International Paper Triennale (Swiss)

ARTWORK TITLE: Korean Pojagi   MATERIAL: Korean hemp   SIZE: 45cm×55cm

ARTIST NAME: Park, Ha-Na
COUNTRY: Korea

CURRICULUM VITAE:
Ph. D candidate, Hongik University, Seoul, Korea
M.F.A., Konstfack University College of Arts, Crafts and Design, Stockholm, Sweden
B.F.A., Konstfack University College of Arts, Crafts and Design, Stockholm, Sweden
Many times of solo & group exhibition in Korea, Japan, China and Sweden
Present: Lecturer at Hongik University

ARTWORK TITLE: Old Tree   MATERIAL: wool, copper, stainless steel mesh   SIZE: 75cm×95cm

ARTIST NAME: Park, Hye-Yeon
COUNTRY: Korea

CURRICULUM VITAE:
2009.3—2010.5 Assistant Pattern Designer, Hue Design Company
2010.6 Pattern CAD Designer, ENS Apparel Company
2009.2 ESMOD SEOUL (Diplome)
2008 A winner of the participation prize of JoongAng Design Contest
2008 A winner of the Golden Thimble Prize of ESMOD PARIS

ARTWORK TITLE: Kippy's Garden
MATERIAL: naturally dyed from artiodactyla cotton fabric   SIZE: 50cm×120cm

姓名：秦寄岗
国籍：中国

简历：清华大学美术学院染织服装系副教授
从事多年专业教学与研究工作

作品名称：《衔花·散红》　材料：丝绸、植物扎染、丝线　尺寸：45cm×86cm

ARTIST NAME: Rhee, Myung-Soog
COUNTRY: Korea

CURRICULUM VITAE:
The 3rd Myung-Soog Rhee Fiber Arts Exhibition (Seoul, Japan)
The Korean Fiber Arts 100 Artist Series (Seoul)
The Japan-Korean Fiber Arts Exhibition (Korean Cultural Service, Japan)
Invited "Natural Dyed Wrapping Cloth" (Korean Culture Center in LA, America)
Invited "Voice of Korean Textile" (Lodz City, Poland)
Invited Exhibition of Design Messe (Frauen Museum, Germany)
Present: Professor of Department Fashion Design, Konkuk University

ARTWORK TITLE: In the Wind   MATERIAL: ramie fabric, silk yarn   SIZE: 140cm × 140cm

ARTIST NAME: Ryu, Myung-Sook
COUNTRY: Korea

CURRICULUM VITAE:
B.F.A., Industrial Design of Chosun Unversity
M.F.A., Graduate School of Ehwa Womens University, Seoul
Adjunct Professor, Traditional Costume, Baewha Women's University, Seoul
Certificate of Natural Dyeing Instructor

ARTWORK TITLE: Indigo Blue Sea... Pear Floral   MATERIAL: cotton yarn (indigo, kochinil, onion)   SIZE: 80cm×150cm

姓名：任晟萱
国籍：中国

简历：2008—2009年 获清华之友柒牌一等奖学金
2009—2010年 获国家励志一等奖学金
2010年 作品《漾》入围海宁经编概念服
2010年 作品《风彝情》获"彝族刺绣创意设计"银奖
2011年 作品《家园》获"全国纺织品设计大赛"优秀奖
2011年 作品《天空之城》获"张謇杯国际纺织品设计大赛"创意设计组金奖
2011年 作品《音乐之声》获"张謇杯国际纺织品设计大赛"创意设计组银奖

作品名称：《原味生活》 材料：棉、麻、靛蓝 尺寸：55cm×450cm

姓名：闪秀桂
国籍：中国

简历：闪秀桂，女，教授

作品名称：南阳汉画《嫦娥奔月》　　材料：细条绒　　尺寸：48cm×110cm

姓名：沈晓平
国籍：中国

简历：天津美术学院设计艺术学院副教授
2007—2008年 新西兰尤尼泰克理工学院和奥克兰商学院访问学者及研修
2008年 设计作品获"亚洲联盟超越设计展"最佳作品奖
2008年 作品参展"中日纤维艺术交流展"
2011年 作品参展"2011国际拼布艺术展"

作品名称：《天际系列1号日象》、《天际系列2号月象》　　材料：雨露麻、混合染料　　尺寸：185cm×66cm

姓名：石历丽
国籍：中国

简历：服装专业硕士毕业，西安美术学院服装系副教授、服装设计与工程教研室主任，中国服装设计师协会会员。在《装饰》、《美术观察》等核心刊物发表多篇论文和作品，作品多次参加比赛与展览。

作品名称：《涟漪》　　**材料**：涤棉布、紫甘蓝汁、蔬果汁　　**尺寸**：75cm×152cm

ARTIST NAME: Sohn, Hee-Soon
COUNTRY: Korea

CURRICULUM VITAE:
2011 International Textiles & Costume Exhibition 2011, "Heritage Textiles and Costume",
ITB (Inonesia)/ CCA(Korea)/Ars Textrina (UK), Institut Teknologi Bandung, Inonesia
2011 10th International Fashion Exhibitionby Invitation -Timeless Eco,The Costume Culture Association/Art Textile, Sookmyung Women's University, Seoul, Korea
2009 Korea-Thailand International Exhibition & Fashion Show-Asia
Forever, In Commemoration of 50th Anniversary Korea-Thailand Diplomatic
Relations, Silpakorn University, Bangkok, Thailand
2008 Korea-China International Invited Fashion Exhibition-Art & Asian Beauty, Tsinghua University, Beijing, China
2007 KSFB & HKPU International Fashion Exhibition, The Hong Kong Polytechnic University, Hong Kong

ARTWORK TITLE: BLATHA's Dream 2012
MATERIAL: silk fabric dyed with natural dyes   SIZE: 80cm×135cm

姓名：山崎和树
国籍：日本

简历：博士，日本东北艺术工科大学美术科副教授、染色工艺家、草木工房主宰
1982年 在父亲山崎青树（群马县指定重要无形文化财保持者）的指导下开始植物染色的研究
2002年 信州大学工学系研究科博士后期课程修了学术博士
2008年 东北艺术工科大学美术科染织专业副教授

作品名称：《无题》　　材料：真丝　　尺寸：3.8m×0.21m

姓名：单夏丽
国籍：中国

简历：作品《蝶语》 获得"2011年中国国际家用纺织品创意设计大赛" 创意设计银奖
　　　地毯作品《晨曦中的风景》 获得"2011'鲁绣杯'中国大学生家用纺织品创意设计大赛"金奖
　　　作品《梦游马戏团》 入选"2011年全国纺织品设计大赛暨国际理论研讨会"，获得银奖
　　　作品《蜕变》 入选"2011年全国纺织品设计大赛暨国际理论研讨会"，获得优秀奖
　　　2008年考入清华大学美术学院染织服装艺术设计系

作品名称：《汐》　　材料：绡、玻璃珍珠　　尺寸：5m×1.5m

姓名：田青
国籍：中国

简历：1953年生于北京，1982年毕业于中央工艺美术学院染织美术系并留校任教。现任清华大学美术学院染织服装艺术设计系教授、博士生导师。多年来一直从事染织艺术设计与教学工作，作品曾多次在国内外展出并获奖。系全国纺织教育学会理事；中国家纺协会设计师分会副主席；中国流行色协会理事；中国科学技术协会决策咨询专家库专家等。

作品名称：《上善若水》　　材料：桑蚕丝　　尺寸：35cm×600cm

姓名：吴越齐
国籍：中国

简历：2009年至今 广州美术学院工业设计学院纤维艺术设计工作室教师
2009年 清华大学美术学院染织服装艺术设计系染织MA
2005年 清华大学美术学院染织服装艺术设计系染织BA
作品曾参加英国曼彻斯特都会大学交流、亚洲纤维艺术作品展等。主要从事印染和毡艺的研究及教学工作。

作品名称：《水的遐想》　　材料：真丝　　尺寸：200 cm×50cm×3cm

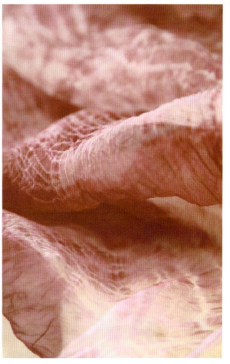

姓名：王斌
国籍：中国

简历：1979年7月生于内蒙古。2003年毕业于天津美术学院服装染织系染织设计专业，学士；2003—2006年任职于深圳富安娜家居用品股份有限公司研发部；2008年毕业于韩国东亚大学研究生院纤维造型设计专业，硕士；2008年任职于山东工艺美术学院现代手工艺术学院，讲师。中国工艺美术学会纤维艺术专业委员会会员。

作品名称：《消失的记忆》　　材料：真丝　　尺寸：90cm×120cm

姓名： 吴元新
国籍： 中国

简历： 中国工艺美术大师，中国民间文艺家协会副主席，国家级非物质文化遗产传承人，中国民间文化杰出传承人，中国艺术研究院客座研究员，享受国务院政府特殊津贴专家，荣获第三届全国中青年德艺双馨文艺工作者。现任南通大学蓝印花布艺术研究所所长、南通蓝印花布博物馆馆长。

作品名称：《喜相逢》　　**材料：** 棉、蓝染

姓名：吴波
国籍：中国

简历：清华大学美术学院副教授
作品多次参加"全国美展艺术设计展"、"艺术与科学国际作品展"、"亚洲纤维艺术展"、"国际纤维艺术双年展"等中、外展览。在国内、国际赛事中获得多项金、银奖，并荣获"国际最佳青年服装设计师"称号。

作品名称：《暮》　材料：毛毡　尺寸：76cm×200cm

姓名：朱小珊
国籍：中国

简历：清华大学美术学院染织服装艺术设计系副教授
作品多次参加"全国美展艺术设计展"、"艺术与科学国际作品展"、"亚洲纤维艺术展"、"国际纤维艺术双年展"等中、外展览。
曾发表、出版《纸上的游戏》、《衣服中的情感》、《服装设计基础》、《服装配饰剪裁教程》、《艺术设计赏析》、《服装工艺基础》等论文和教材。

作品名称：《日》　　材料：毛毡　　尺寸：76cm×200cm

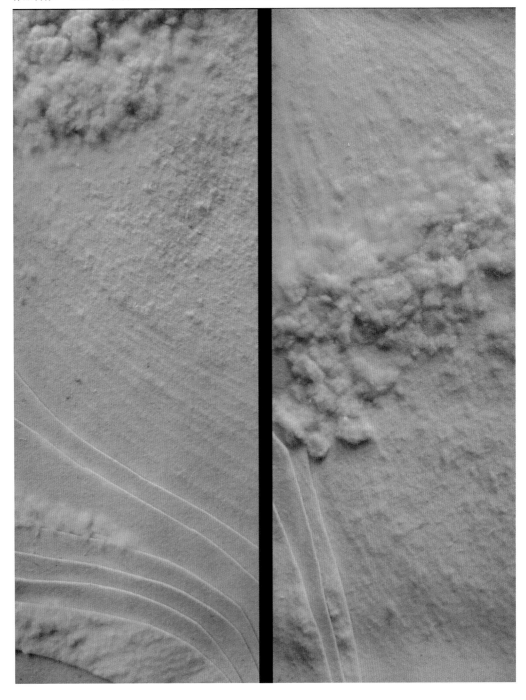

姓名：王懿龙
国籍：中国

简历：清华大学美术学院染织服装艺术设计系本科四年级学生。
在大学学习期间，一直致力于探索本土性与当代性在设计上的结合，热爱中国传统文化，渴望扩大中国传统文化的影响力，设计作品多次在国内和国际性比赛中获奖。

作品名称：《生生不息》　　材料：绡　　尺寸：450cm×1.2cm

姓名：许韫智
国籍：中国

简历：2005—2009年就读于清华美院装潢艺术设计系，2009年考入清华美院染织艺术设计系。2010年作品《和》参展"第七届亚洲纤维艺术展"。2011年针织拼布靠垫入选《2011年第十一届全国纺织品设计大赛暨国际理论研讨会作品集》。爱好编织，热爱染织设计。希望在纺织品设计中融入更宽阔的功能性的创新概念。

作品名称：《棍·茶》　　材料：丝、棉毛　　尺寸：45cm×45cm

姓名：萧静芬
国籍：中国

简历：现任职于台湾工艺研究发展中心技术组染织工坊，并于亚洲大学数位媒体设计研究所时尚设计组进修中，学习及制作植物蓝靛染色工艺十余年。

作品名称：《昙花》　　材料：棉布、植物蓝靛染料　　尺寸：50cm×75cm

姓名：徐秋宜
国籍：中国

简历：一位以艺术应用的研究实验者自许的创作者，是介乎设计师和市场需要及梦想的执业人。在台湾美术系西画组毕业，曾获台阳美展雕塑组第三名。旅法7年，其间获工业局奖学金，并数次获国际服装设计首奖，法国媒体称她的服装线条有如感性的动态雕塑。

她擅长人体素描、雕塑、打版与空间装置，拥有法国高级时装公会学校立裁专门技术与法国知名学府国立巴黎高等装饰艺术学院（ENSAD）服装设计系硕士文凭，创作领域结合艺术与时尚，自1996年迄今分别在中国大陆、香港、台湾，以及法国共举办过多场大型设计师联展与个人服装展，目前更致力于"生态生活、时尚与艺术"及多元媒材后加工工艺的创作应用与发展。

作品名称：《剪衣》　　**材料**：丝、椰丝、羊毛

姓名：杨颐
国籍：中国

简历：天津工业大学设计艺术学院艺术硕士。曾到法国ESMOD国际服装学院深造，并在法国及国内品牌服装公司从事产品开发工作。现任广州美术学院工业设计学院讲师，专注家纺产品、面料设计与开发的教学事业，并跨界游走于服装、家纺及家居软装设计之间。

作品名称：《花·雪》　　材料：毛毡、麻、丝　　尺寸：180cm×120cm

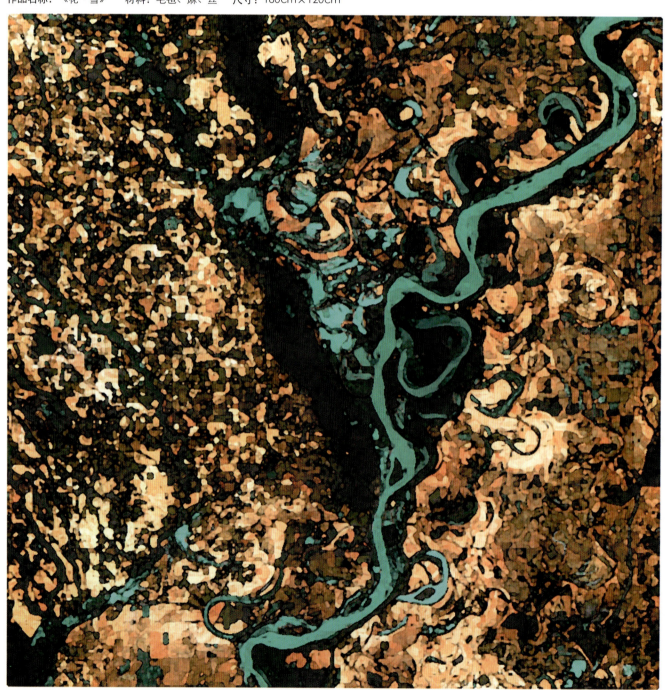

姓名：杨建军
国籍：中国

简历：清华大学美术学院教师。1998年毕业于中央工艺美术学院（现清华大学美术学院）染织服装艺术设计系，获硕士学位并留校任教。2008—2009年作为中国政府派遣研究员赴日本东京艺术大学客座研究天然染色材料与工艺技术。

作品名称：《如是观》　　材料：丝（面料），洋葱皮（染料），草木灰、钛、铝、铜、铁（媒染剂）　　尺寸：114cm×600cm

姓名：于婷婷
国籍：中国

简历：2009年 考入清华大学美术学院染织服装艺术设计系
　　　2010年 作品《和》参展"第七届亚洲纤维艺术展"
　　　2011年 作品《蓝·源》参展"2011国际拼布艺术展"

作品名称：《涓流》　　材料：乔其

姓名：杨芳
国籍：中国

简历：1985年9月出生。毕业于清华大学美术学院染织服装艺术设计系，硕士研究生。目前任爱慕内衣有限公司材料设计工作室设计师，从事爱慕旗下各品牌的印花设计、提花设计，及流行趋势预测等工作。

作品名称：《月白·飞白》　　材料：真丝顺纡、真丝绡　　尺寸：160cm×40cm

ARTIST NAME: Yoo, Ja-Hyung
COUNTRY: Korea

CURRICULUM VITAE:
Graduated from School of Department Visual Design, SEOUL Arts College (1980)
Group Exhibition of The Chang-Yeom Association (1983—1996)
International Exhibition of Natural Dye For Invitional & Exchange (2008)
Prize at 4th Yongsan International Art Exhibition (2008)
Natural-dyeing exhibition "An Outing" 2011(Artshop Gallery of Seoul Arts Center)
Present: Fiber artist

ARTWORK TITLE: Harmoney   **MATERIAL:** silk, sapanwood tree, gardenia (alum of fixatives)   **SIZE:** 240cm×110cm

ARTIST NAME: Yoon, Jae-Shim
COUNTRY: Korea

CURRICULUM VITAE:
2003 M.A. Graduated from School of Art & Design Sangmyung University Korea
2012 Ph. D. in Arts Graduated from School of Sangmyung University Korea
Present: lecturer at Hanbat National University

ARTWORK TITLE: Start of Light
MATERIAL: Korean paper, chacoal, glittering powder   SIZE: 50cm × 200cm

姓名：张洁
国籍：中国

简历：2007—2011年就读于广州美术学院，在染织艺术设计教研室（即现在的工业设计学院纤维艺术设计工作室）进行纤维方向的学习。

姓名：朱建重
国籍：中国

简历：2007—2011年就读于广州美术学院，在染织艺术设计教研室（即现在的工业设计学院纤维艺术设计工作室）进行纤维方向的学习。

作品名称：《复刻自然》　　材料：绡、丝麻　　尺寸：(150cm×150cm)×3

姓名：赵莹
国籍：中国

简历：鲁迅美术学院染织服装艺术设计系染织专业研究生
2008年 获全国家纺设计大赛银奖
2009年 获评"辽宁省优秀毕业生"
2011年 作品被选入"第八届亚洲纤维艺术展"参展。获首届"中国纤维艺术展"优秀奖
获"中国国际面料设计大赛"特别奖，此奖获评2011年度沈阳高校十大新闻

作品名称：《流年》　材料：麻布、线、丙烯、木珠　尺寸：120cm×240cm

姓名：郑晓红
国籍：中国

简历：1999年至今 任日本染织设计家协会（TDA）会员
2009年 任中国美术家协会会员

作品名称：《鱼》　材料：麻布　尺寸：110cm×160cm

姓名：张树新
国籍：中国

简历：清华大学美术学院染织服装艺术设计系副教授、硕士研究生导师
北京工艺美术学会理事、中国工艺美术学会纤维艺术专业委员会理事
编织壁挂《烛光·汶川》2008年11月入选"'从洛桑到北京'——第五届国际纤维艺术双年展"
编织壁挂《后青花时代》2008年入选"'精工植物'——第四届中国现代手工艺术学院展"
编织壁挂《流光》2009年入选"2009中日韩美国际交流展示会"
编织壁挂《东学西渐》2010年3月入选"第七届亚洲纤维艺术作品展"

作品名称：《夏日》　　材料：植物染（蓝靛、栀子、苏木、核桃皮）羊毛　　尺寸：75cm×80cm

姓名：张宝华
国籍：中国

简历：清华大学美术学院染织服装艺术设计系副主任、副教授、硕士生导师
中华全国工商业联合会纺织服装商会专家委员会委员
中国家用纺织品行业协会设计师分会副会长
中国流行色协会色彩教育委员会委员
NCS (Natural Color System) 中国地区特约色彩专家
1990年 毕业于中央工艺美术学院染织艺术设计专业，获学士学位
2003年 毕业于香港理工大学纺织品及服装设计专业，获硕士学位

作品名称：《竹》　　材料：麻和化纤交织面料　　尺寸：142cm×55cm

姓名：张靖婕
国籍：中国

简历：山东工艺美术学院教师，从事染织艺术设计的教学工作，主要研究方向为染织材料的创新应用。
主要作品有《简单生活》、《本原》、《墨·时代》等。作品多次参加展览并获奖。

作品名称：《重生》　　材料：蚕茧、丝　　尺寸：100cm×300cm

姓名：朱微婷
国籍：中国

简历：清华大学染织服装系染织设计专业大四学生，擅长手绘和染织工艺，造型能力强，熟练运用Photoshop、Coreldraw等常用绘图软件。思维活跃，眼界开阔，性情温和，个性乖巧细心，热爱中国文化，对世界各地的民族文化充满求知欲。曾获中国领带"'名城杯'第三届全国丝品花型设计大赛"优秀奖等奖项。

作品名称：《书生》　　材料：绡　　尺寸：4m×1.2m

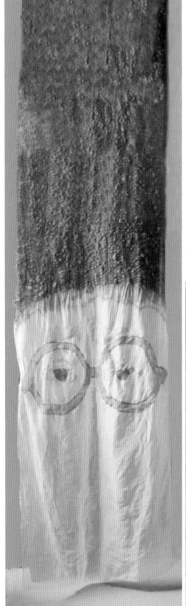

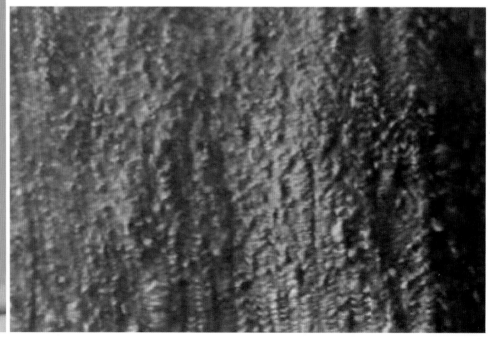

姓名：朱医乐
国籍：中国

简历：天津美术学院副教授
中国美术家协会天津分会会员
中国工艺美术学会会员
中国纤维艺术学会理事
2011年 作品《塑造》入选"2011国际拼布艺术展传承与创新拼布作品展"。
2011年 论文《纤维艺术课程教学感悟材料、构思、表现》发表于《第十一届全国纺织品设计大赛暨国际理论研讨会论文集》中。

作品名称：《叠加》　　材料：麻纤维植物染色　　尺寸：40cm×60cm

姓名：张红娟
国籍：中国

简历：清华大学美术学院染织服装艺术设计系讲师
作品多次参加国内外纺织艺术设计大赛，亚洲纤维艺术展、国际纤维艺术双年展等中、外展览
在国内、国际赛事中获多项金、银奖

作品名称：《夕》　　材料：毛料、黄连　　尺寸：50cm×100cm

姓名：杨文斌
国籍：中国

简历：苗族，1942年出生于贵州省雷山县控拜村银匠世家。美术专科毕业，从事西南、华南、中南少数民族服饰研究和教学27年。曾被北京服装学院、凯里学院聘为研究员、兼职教授等职，现为贵州黔东南州方蒿民族工艺品有限公司任植物染色设计师。
著书有《苗族传统蜡染》、《贵州蜡染》等。

作品名称：《蝴蝶妈妈》　　材料：棉布、染色用植物蓝靛浸染　　尺寸：270cm×162cm

姓名：Kim, Kwang Soo
国籍：韩国

简历：1976—2001年 光镇出版社代表，古潭传统染色研究所所长
　　　1989—2009年 釜山茶人联合会讲士
　　　1983年—今 古潭液蓝（Godam indigo）代表
　　　2009年—今 古潭出版社代表
　　　1977年—今 大韩佛教僧家宗监查院长
　　　2011年—今 古潭传统文化会代表
　　　2001年 液蓝（indigo）制造方法和熟成装置出愿(特许 第0420990 获得)

作品名称：靛蓝染　　材料：棉、麻、丝

ARTIST NAME: Reiko Hara
COUNTRY: Japan

CURRICULUM VITAE:
Reiko Hara is both a textile artist and flower artist. She is an observer and appreciator of nature: the balance it presents, its beauty, and the revolution of seasons. Fittingly, her work seeks to harmonize nature, material, and design. Reiko likes when her art surprises viewers or calms them; generally she aims to project positivity on those who see and experience her work.

ARTWORK TITLE: Remains   MATERIAL: cotton   SIZE: 100cm×150cm

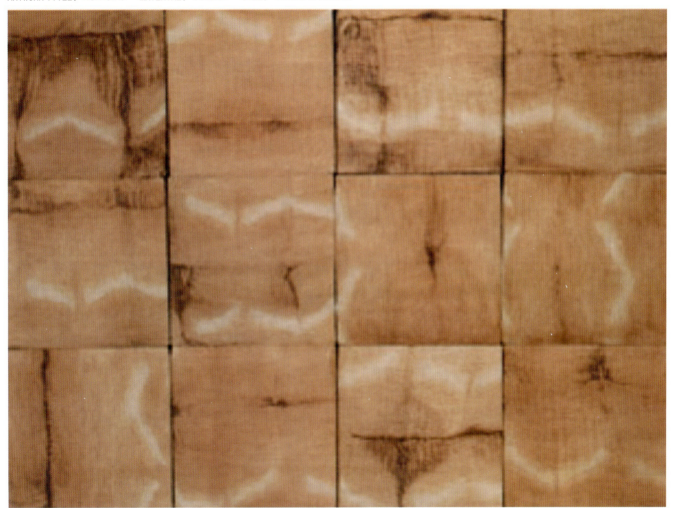

姓名：王晶晶
国籍：中国

简历：2012年 获清华大学美术学院硕士学位
2002年 作品参加"第三届亚洲纤维艺术展"
2009年 作品参加国际纺织品大赛并获奖
2010、2011年 参加"全国纺织品大赛暨国际理论研讨会"。设计作品多次获得国际、全国专业设计大赛奖项。

作品名称：《母与子》　　材料：棉、天然染料　　尺寸：116cm×70cm

姓名：刘玥
国籍：中国

简历：2009年 就读于清华大学美术学院染织服装系
2011年 获得清华之友——光华奖学金

作品名称：《韵》　　材料：丝绸、石榴皮、姜黄　　尺寸：160cm×70cm

姓名：曹宇坤
国籍：中国

简历：2007年 就读于清华大学美术学院染织服装系

作品名称：《茶染》　　材料：丝、麻

ARTIST NAME: Sara Ashford
COUNTRY: America

CURRICULUM VITAE:
I am a lifelong artist living in the foothills in the very Northwest corner of the United States. Using a variety of cloth substrates, I experiment with the entire complex of natural dyeing processes and techniques. Working in small series and following my intuition with each piece, I begin adding any variety of techniques from the natural dye galaxy that seems appropriate to achieve the developing vision. I get in my mind for that particular piece. Being part of the ancient plant dye tradition fills me with Awe.

ARTWORK TITLE: Untitled   MATERIAL: silk   SIZE: 44cm×120cm

姓名：杨锦雁
国籍：中国

简历：2010年 清华大学美术学院硕士研究生毕业
　　　作品《奇色异彩》2010年获"全国纺织品设计大赛" 铜奖
　　　作品《山水之间》系列作品入选2010年"第七届亚洲纤维艺术展"
　　　作品《万物生》入选 2010年"从洛桑到北京——第六届国际纤维艺术双年展"
　　　作品《寻找香巴拉》入选"2011国际拼布艺术展——传承与创新"

作品名称：《阿锦》　　材料：棉布　　尺寸：180cm×60cm

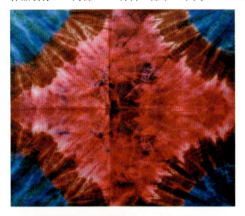

民间植物染作品
Folk Plant Dyeing Works

- 印度
- 非洲
- 广州
- 广西
- 江苏
- 云南
- 浙江
- 新疆
- 贵州
- 海南

拓印 印度（1）

拓印 印度（2）

拓印　印度（3）

拓印 印度（4）

糊染 非洲（1）

糊染 非洲（2）

香云纱　薯莨染　广州　顺德　汉族

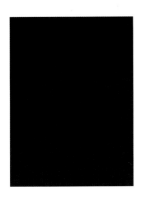

靛蓝、薯莨、牛胶染　广西　那坡　黑衣壮族（1）

靛蓝、薯莨、牛胶染　广西　那坡　黑衣壮族（2）

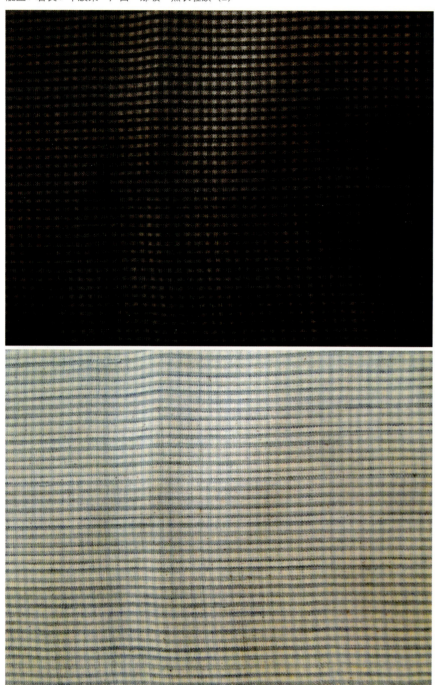

靛蓝染　广西　那坡　黑衣壮族

蓝印花布　靛蓝　"纺织图"壁挂　江苏　南通　汉族

蓝印花布　靛蓝　"飞天"壁挂　江苏　南通　汉族

蓝印花布　靛蓝　"凤戏牡丹"桌布　江苏　南通　汉族

蓝印花布　靛蓝　"年年有余"工艺品系列　江苏　南通　汉族

蓝印花布　靛蓝　江苏　南通　汉族（1）

蓝印花布　靛蓝　江苏　南通　汉族（2）

蓝印花布　靛蓝　江苏　南通　汉族（3）

蓝印花布　靛蓝　江苏　南通　汉族（4）

蓝印花布　靛蓝　江苏　南通　汉族（5）

蓝印花布　靛蓝　江苏　南通　汉族（6）

蓝印花布　靛蓝　花卉动物纹　江苏　南通　汉族

蓝印花布　靛蓝　花卉纹　江苏　南通　汉族

蓝印花布　靛蓝　几何纹　江苏　南通　汉族（1）

蓝印花布　靛蓝　几何纹　江苏　南通　汉族（2）

彩色浇花布　江苏　南通（1）

彩色浇花布　江苏　南通（2）

植物染色织布 江苏 南通（1）

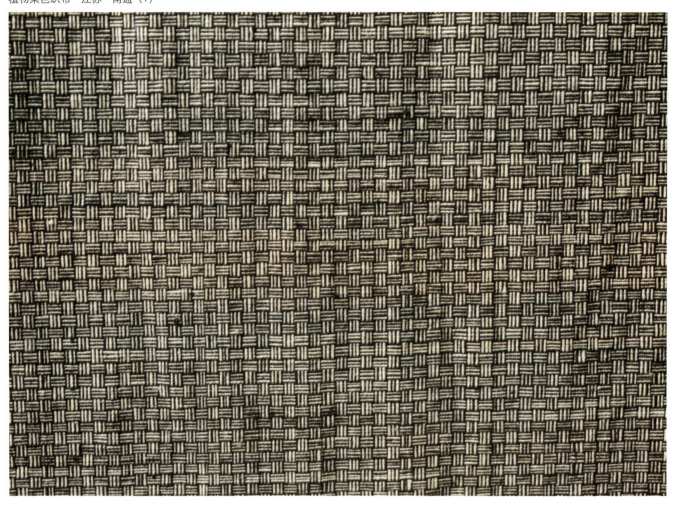

植物染色织布　江苏　南通（2）

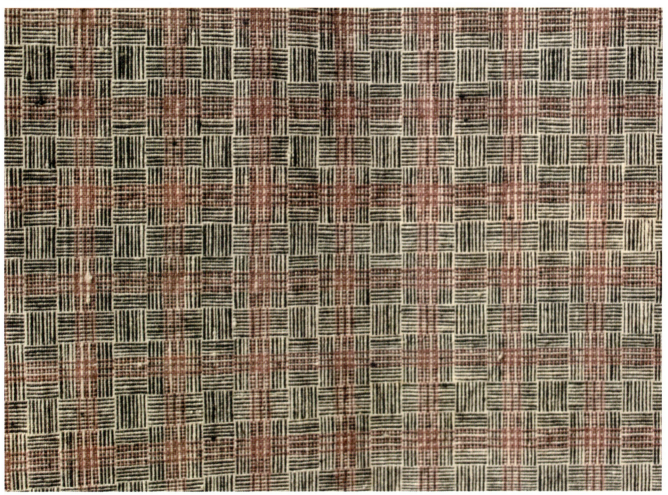

植物染色织布　江苏　南通（3）

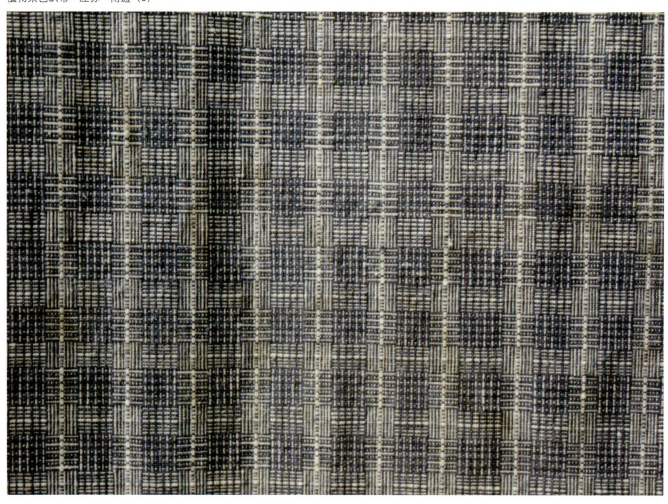

植物染色织布　江苏　南通（4）

扎染 靛蓝 云南 大理 白族 (1)

扎染　靛蓝　云南　大理　白族（2）

扎染 靛蓝 云南 大理 白族（3）

扎染 靛蓝 云南 大理 白族（4）

扎染　靛蓝　云南　大理　白族（5）

扎染　靛蓝　云南　大理　白族（6）

扎染 靛蓝 云南 大理 白族（7）

扎染　靛蓝　云南（1）

扎染 靛蓝 云南（2）

扎染 靛蓝 云南（3）

扎染 云南 大理 白族（3）

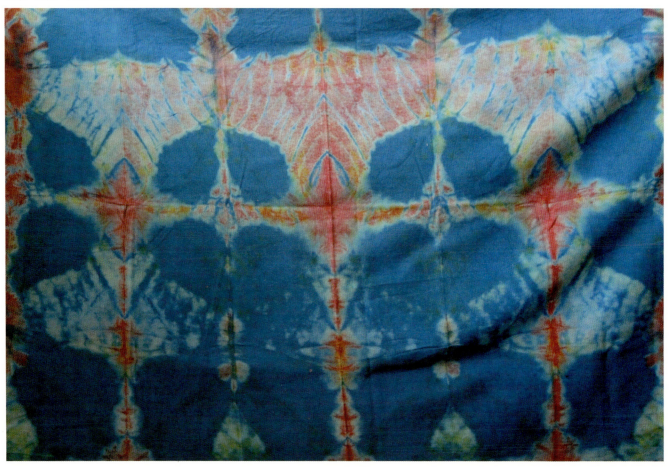

扎染　靛蓝　云南（4）

扎染　靛蓝　云南（5）

夹缬 靛蓝 浙江 温州（1）

夹缬　靛蓝　浙江　温州（2）

夹缬　靛蓝　浙江　温州（3）

夹缬　靛蓝　浙江　温州（4）

夹缬　靛蓝　浙江　温州（5）

夹缬　靛蓝　浙江　温州（6）

夹缬 靛蓝 浙江 温州（7）

夹缬　靛蓝　浙江　温州（8）

夹缬　靛蓝　浙江　温州（9）

夹缬　靛蓝　浙江　温州（10）

夹缬　靛蓝　浙江　温州（11）

艾德丽丝绸　枸杞根、桑树瘤子、生铁渣染色　新疆　和田

艾德丽丝绸　红柳枝、凤仙花、卡列古丽染色　新疆　和田

艾德丽丝绸　红柳枝、茜草、槐花染色　新疆　和田

艾德丽丝丝绸　红柳枝、卡列古丽、茜草染色　新疆　和田

艾德丽丝丝绸　核桃皮、卡列古丽、奥斯曼染色　新疆　和田

蜡染 靛蓝染 广西 苗族

靛蓝染　广西　隆林　苗族

蜡染 靛蓝染 广西

粘膏染 靛蓝染 广西 白裤瑶族

植物染彩色群（清代）贵州 从江县 苗族

蜡染刺绣妇女上衣(民国) 贵州 纳雍 苗族

妇女蜡染上衣（民国） 贵州 镇宁 布依族

妇女蜡染刺绣上衣（民国） 贵州 贵定 苗族

蜡染被面（清末） 贵州 三都 苗族

蜡染被面　贵州　丹寨　苗族

蜡染 靛蓝染 贵州 苗族（1）

蜡染 靛蓝染 贵州 苗族（2）

蜡染被面(清代) 贵州 惠水 苗族

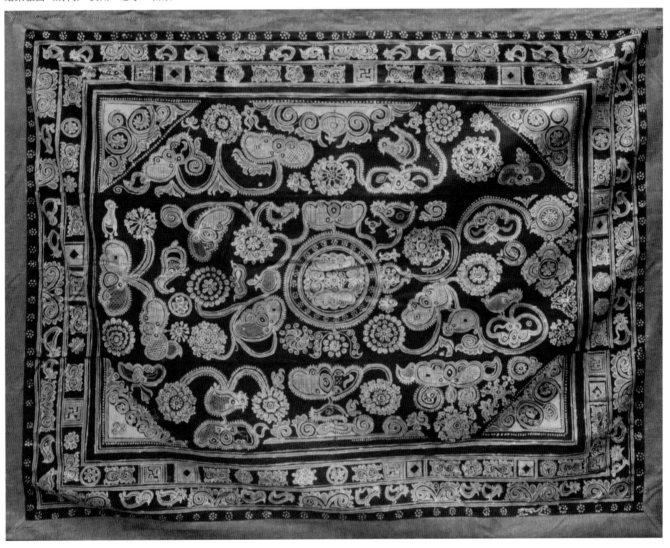

衣袖片（民国）贵州　黔西　苗族

蜡染围腰 贵州 榕江

蜡染刺绣衣袖（民国）贵州 黄平革家

植物彩色蜡染背带片（民国） 贵州 黔西 苗族

蜡染植物染色拼布儿童包巾（清代） 贵州 惠水 布依族

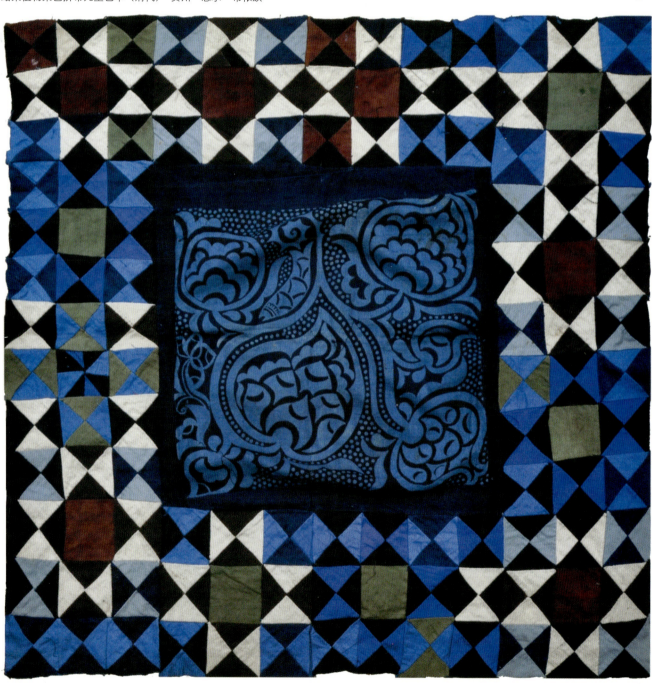

织金蜡染方巾（现代） 贵州

蜡染祭鼓幡　贵州　榕江县　高排苗族

蜡染 贵州

蜡染　靛蓝、金沟藤、黄栀子等染色　贵州

蜡染 靛蓝染 贵州

蜡染刺绣妇女头帕(民国) 海南 苗族

素纱禅衣（复制品）说明：
　　1987年，在国家文物局和湖南长沙马王堆博物馆的要求下，南京云锦研究所承担了复制长沙马王堆出土的汉代"素纱禅衣"工作。根据复制任务的需要，南京云锦研究所成立了古代天然染料研究小组，经过3年多的调研、资料收集、整理和试验，于1991年成功的复制了"素纱禅衣"，还原了古代染色法，解决了素纱禅衣中的"绒圈锦"和"素纱"古代染色的难题。

素纱禅衣（长沙马王堆出土，南京云锦研究所复制）

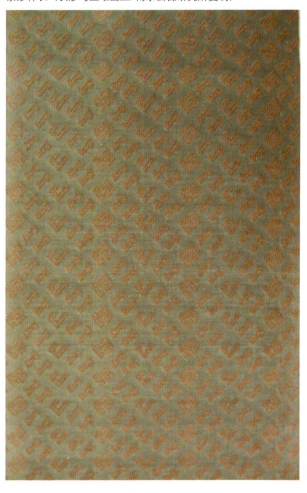

纺织艺术设计
TEXTILE DESIGN

2012年第十二届全国纺织品设计大赛暨国际理论研讨会
12TH CHINA TEXTILE DESIGN COMPETITION & INTERNATIONAL CONFERENCE 2012

2012年国际植物染艺术设计大展暨理论研讨会——传承与创新
INTERNATIONAL PLANT DYEING ART EXHIBITION & CONFERENCE—INHERITANCE & INNOVATION 2012

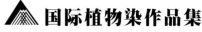

国际植物染作品集
WORKS COLLECTION OF INTERNATIONAL PLANT DYEING

主办单位：	清华大学艺术与科学研究中心
联合举办：	中国家用纺织品行业协会
	中国纺织服装教育学会
	中国流行色协会
	中国工艺美术协会
	清华大学美术学院
承办单位：	清华大学美术学院染织服装艺术设计系
组委会：	全国纺织品设计大赛暨国际理论研讨会组委会成员（按姓氏笔画排序）
	王　利　天津美术学院　教授
	王庆珍　鲁迅美术学院　教授
	田　青　清华大学美术学院　教授
	朱尽晖　西安美术学院　副教授
	李加林　浙江理工大学　教授
	吴海燕　中国美术学院　教授
	余　强　四川美术学院　教授
	张　莉　西安美术学院　教授
	张　毅　南通大学　副教授
	张宝华　清华大学美术学院　副教授
	张树新　清华大学美术学院　副教授
	陈　立　清华大学美术学院　副教授
	郑晓红　中国人民大学　副教授
	秦岱华　清华大学美术学院　副教授
	贾京生　清华大学美术学院　教授
	龚建培　南京艺术学院　教授
	崔　唯　北京服装学院　教授
	韩丽英　天津美术学院　教授
	霍　康　广州美术学院　副教授

参展单位：

韩国植物染艺术家
韩国古潭传统文化会
Korea Traditional Artshop & Center
Korean Traditional Culture of Art Union
Korea Arts and Crafts "Jung-Whi" Research Center
The Association of Korean Knitting Design
Metropolitan Architecture Company, Korea
日本植物染艺术家
印度尼西亚植物染艺术家
马来西亚植物染艺术家
美国植物染艺术家
印度植物染艺术家
非洲植物染艺术家
中国台湾植物染艺术家
台湾工艺研究发展中心技术组染织工坊
台湾天染工坊
台湾卓也小屋手工坊
Taiwan Art Quilt Society
中国民间植物染艺术家
"ONE SUN" 刺绣工艺研究室
原野天然染料染色工作室
中国艺术研究院
南京云锦研究院
南通蓝印花布博物馆
南通三友民间艺术研究所
江苏常州云卿纺织品有限公司
贵州黔东南州方蒿民族工艺品有限公司
爱慕内衣有限公司
Ewha Womans University, Korea
Kyungwon University, Korea
Seoul National University, Korea
Dongduk Women's University, Korea
Hanyang Women's University, Korea
Hanbat National University, Korea
Hanyang University, Korea
Sangmyung University, Korea
Sookmyung MYUNG Women's University, Korea
Sookmyung Women's University, Korea
Suwon Women's College, Korea
Kyung Hee University, Korea
Gyeongnam National University, Korea
Hongik University, Korea
Chung-Ang University, Korea
Konkuk University, Korea
Baewha Women's University, Seoul, Korea
Tama Art University
Institute of Technology, Bandung, Indonesia
日本东北艺术工科大学
日本多摩美术大学
台湾亚洲大学
清华大学美术学院
中国美术学院
鲁迅美术学院
广州美术学院
南京艺术学院
四川美术学院
西安美术学院
浙江理工大学
北京服装学院
天津美术学院
中国人民大学
云南民族大学
西安工程大学
南通大学
山东工艺美术学院
青岛大学
南阳理工学院
中国防卫科技学院
东北大学
四川理工学院
苏州大学

（排名不分先后）

活动内容与时间：

国际理论研讨会：2012年3月26日 下午1：30—5：00
　　　　　　　　2012年3月27日 上午9：00—12：00
纺织设计作品展：2012年3月26日— 4月2日
国际植物染艺术展：2012年3月26日— 4月2日

地　　点： 清华大学美术学院

赞助单位： 意大利 GUCCI 集团
　　　　　　 中国建筑工业出版社
　　　　　　 福州迪捷特数码科技有限公司
　　　　　　 江苏常州依丽雅斯纺织品有限公司
　　　　　　 广东东莞市天龙化工实业有限公司

标识设计： 田旭桐

策　　划： 田青

策　　展： 杨冬江

清华大学艺术与科学研究中心
2012年第十二届全国纺织品设计大赛暨国际理论研讨会组委会
中国家用纺织品行业协会
中国纺织服装教育学会
中国工艺美术协会
中国流行色协会
清华大学美术学院染织服装艺术设计系
2012年3月